MULTILAYER SWITCHING TECHNOLOGY

多层交换技术

实践篇

汪双顶 袁晖 史振华／主编
王明昊 周桐 赵景／副主编
王隆杰 安淑梅 张璐琦／主审

人民邮电出版社
北京

图书在版编目（CIP）数据

多层交换技术. 实践篇 / 汪双顶, 袁晖, 史振华主编. -- 北京：人民邮电出版社, 2019.11
锐捷网络学院系列教程
ISBN 978-7-115-51736-4

Ⅰ. ①多… Ⅱ. ①汪… ②袁… ③史… Ⅲ. ①网络交换—教材 Ⅳ. ①TP393

中国版本图书馆CIP数据核字(2019)第164756号

内 容 提 要

本书为《多层交换技术》配套实践指导用书，内容涉及园区网工程施工中的多层交换技术，包括 Private VLAN 技术、Super VLAN 技术、MSTP 与 VRRP 技术、RLDP 技术、接口聚合 AP 技术、交换网络安全、VSU 技术以及 WLAN 技术等，通过对这些多层交换技术的学习，读者可掌握保障交换网络可靠性的方法。

本书可作为计算机及相关专业学生学习多层交换组网技术的实践指导用书，也可作为网络工程师及相关技术人员了解园区网多层交换组网技术的参考书。

◆ 主　编　汪双顶　袁　晖　史振华
　 副 主 编　王明昊　周　桐　赵　景
　 主　审　王隆杰　安淑梅　张璐琦
　 责任编辑　左仲海
　 责任印制　王　郁　马振武

◆ 人民邮电出版社出版发行　北京市丰台区成寿寺路 11 号
　 邮编　100164　电子邮件　315@ptpress.com.cn
　 网址　https://www.ptpress.com.cn
　 北京天宇星印刷厂印刷

◆ 开本：787×1092　1/16
　 印张：15　　　　　　　2019 年 11 月第 1 版
　 字数：359 千字　　　　2024 年 9 月北京第 6 次印刷

定价：49.80 元

读者服务热线：(010)81055256　印装质量热线：(010)81055316
反盗版热线：(010)81055315
广告经营许可证：京东市监广登字 20170147 号

前 言 FOREWORD

《多层交换技术（实践篇）》由网络数通厂商数百份园区网工程项目中筛选出的37份具有典型特征的多层交换网络施工文档整理汇编而成。编者希望把这些来自企业的真实园区网施工文档，引入大中专院校日常教学中，使在校生按照企业真实的工作过程完成项目实训任务，掌握园区网中使用到的多层交换技术，了解企业园区网络工程项目实施过程，以便在学习的过程中积累工作经验，在实际工作中恰当地运用这些技术，解决遇到的各种问题。

全书的每一个多层交换工程项目，都从生活中的应用需求出发，以企业真实组网案例为依托，描述多层交换技术在园区网中的应用场景，通过阐述需求、绘制拓扑、规划地址、选型设备、施工及网络测试的过程，使读者掌握园区网构建中应用到的需求分析、网络规划、网络施工、网络测试、网络排障等相关技术。

为更好地实施这些工程项目，实验室需准备好相应硬件设施，包括二层交换机、三层交换机、模块化路由器、万兆模块（可选）、配套线缆（可选）、无线接入AP、无线交换机，以及若干台测试计算机和网线（或制作工具）。如果缺乏相应的硬件环境，也可以使用模拟器软件完成相关的实训项目。

虽然本书选择的工程项目都来自特定厂商的数通项目，但在规划中力求知识诠释和技术选择具有业内通用性，遵循行业内通用技术标准，如设备功能描述、接口标准、技术诠释、命令语法、操作规程和拓扑绘制方法，都使用业内标准术语。

本课程对应的"1+X"证书是资深网络工程师的职业资格认证。学习本书内容后可以参加数通厂商职业资格认证考试，获取证书证明认证者了解协议，熟悉硬件，具有解决网络疑难问题的能力。

本书的开发团队主要由数通厂商工程师和院校的一线专业教师组成，他们将多年来在各自工作领域中积累的网络技术教学和工作经验，以及对网络技术的深刻理解，诠释成本书实践内容。其中，王隆杰教授来自深圳职业技术学院、袁晖教授来自深圳信息职业技术学院，史振华教授来自绍兴职业技术学院，王明昊教授来自大连职业技术学院，周桐教授来自重庆工程职业技术学院，赵景教授来自许昌职业技术学院。作为业内的教学名师，他们多年来工作在教学一线，在国家职业技能竞赛方面具有丰富的大赛训练经验。为完成本课程的开发任务，他们按照院校资深网络工程师人才素质培养的要求，主导全书的项目规划、体例设计，并承担相关文档的修订、实验验证及测试任务，使本书具有通用性。同时，按照技术实施难易度循序渐进地组织规划，以便在大中专院校落地实施。

以汪双顶、安淑梅、张璐琦等为代表的厂商工程师团队，积极发挥了他们在数通厂商拥有的园区网项目和技术资源优势，筛选来自企业的工程项目文档和最新的行业技术解决方案，完成技术和工作场景对接，按照企业中项目实施过程完成项目方案设计，把网络行业最新的技术引入本书，保证技术和市场同步，课程和行业发展一致。

此外，在编写本书过程中，编者还得到了其他一线教师、技术工程师、产品经理的大力支持。他们多年来积累的工程一线的实践经验，为本书的真实性、专业性以及教学实施便利性提供了有力的支持。

本书规划、编辑的过程历经近三年，前后经过多轮修订，但由于课程组人员水平有限，疏漏之处在所难免，敬请广大读者指正。相关教学资源获取请发送邮件至410395381@qq.com 邮箱索取。也可直接访问人邮教育社区（www.ryjiaoyu.com），输入本书名称查询本书资源。

创新网络教材编写委员会
2020 年 4 月于北京

使 用 说 明

为帮助读者全面理解多层交换技术细节，建立直观组网概念，在全书关键技术解释和工程实施中，会涉及一些网络专业术语和词汇，为方便读者今后在工作中应用，全书采用业界标准的技术和图形绘制方案。

以下为本书中所使用的图标示例。

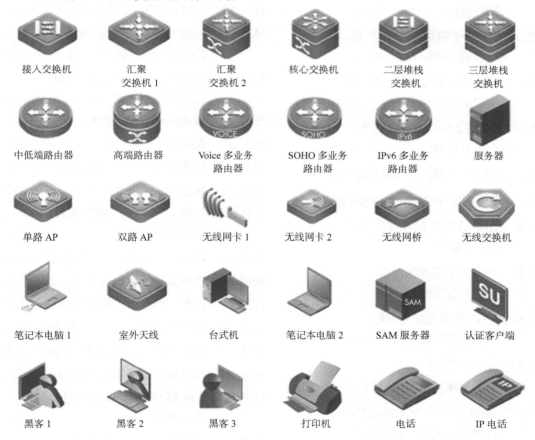

目录 CONTENTS

任务 1	配置跨交换机 VLAN 之间的 SVI 路由	1
任务 2	单臂路由技术实现不同 VLAN 之间通信	6
任务 3	修剪骨干链路 VLAN 广播，优化传输效率	10
任务 4	配置 PVLAN，实现 VLAN 内部隔离	13
任务 5	使用 Super VLAN 技术优化 IP 地址	18
任务 6	使用 STP 技术保障网络健壮稳定	22
任务 7	使用 RSTP 技术加快网络收敛	26
任务 8	使用 MSTP 技术实现网络负载均衡	32
任务 9	VRRP 实现单备份组网络出口冗余 1	40
任务 10	VRRP 实现单备份组网络出口冗余 2	44
任务 11	VRRP 实现多备份组冗余网络出口	48
任务 12	使用 VRRP 技术实现核心网络冗余备份（基于 SVI）	52
任务 13	使用 MSTP+VRRP+AP 技术保障核心网络稳健运行	59
任务 14	配置交换端口二层聚合	81
任务 15	配置三层端口静态聚合	85
任务 16	配置三层端口动态聚合	88
任务 17	配置计算机自动获取 IP 地址	93
任务 18	配置不同子网中设备 DHCP 自动获取地址	99
任务 19	配置交换机 DHCP Relay	105
任务 20	配置 DHCP Snooping 保障 DHCP 服务器的安全性	109
任务 21	使用 RLDP 技术快速检测以太网链路故障	112
任务 22	实施环形 VSU 技术	116
任务 23	配置线形 VSU 技术	124
任务 24	配置 VSU 综合应用案例	129
任务 25	排除 VSU 故障	137
任务 26	使用 NFPP 技术保护交换机 CPU 不受攻击	143
任务 27	IP Source Guard 防范攻击	149
任务 28	保护交换机端口安全	153
任务 29	多对一镜像保护网络安全	157
任务 30	配置交换网络中的一对多镜像	160
任务 31	配置交换机基于流量的镜像安全	163
任务 32	配置接入交换机保护端口	166
任务 33	Fat AP 组建会议室临时放装无线	169
任务 34	组建单核心无线校园网（Fit AP+AC）	172
任务 35	组建单核心无线校园网（Fit AP+SW+AC）	177
任务 36	综合实训 1	183
任务 37	综合实训 2	208

任务 1 配置跨交换机 VLAN 之间的 SVI 路由

【任务描述】

某网络公司为了加强公司信息化建设，组建了互连互通的公司内部办公网络。为避免办公网络中产生的广播干扰，通过 VLAN（虚拟局域网）技术按照部门划分出了不同的部门网络，一个部门一个 VLAN，有效地隔离了部门之间的广播和冲突。但由于不同的 VLAN之间不能通信，因此希望通过在交换机上实施交换虚拟接口（SVI）技术，实现不同部门之间的网络通信。

【任务目标】

配置 SVI 技术，实现 VLAN 之间的路由，实现 VLAN 之间的三层数据通信。

【组网拓扑】

图 1-1 所示的网络拓扑为某网络公司部门网络连接场景，按部门划分出多个 VLAN，实施 SVI 技术，实现了不同部门之间的网络连通。

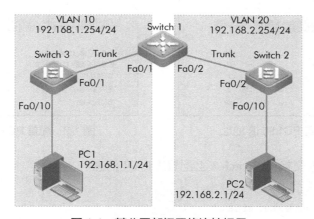

图 1-1 某公司部门网络连接场景

（备注：具体的设备连接接口信息可根据实际情况决定）

【设备清单】

二层交换机（两台），三层交换机（1 台），网线（若干），PC（若干）。

注意：需要说明的是，后续拓扑图中出现的"接入交换机""二层堆栈交换机"图标均属于二层交换机；出现的"固化汇聚交换机""模块化汇聚交换机""核心交换机""三层堆栈交换机"图标均属于三层交换机。因为应用方案不同，有应用细分。为方便使用，后续

多层交换技术（实践篇）

没有特殊说明，分别使用二层交换机、三层交换机名称简化描述，方便开展实训操作。

【关键技术】

VLAN 是一组逻辑上的设备构成网络，这些设备并不受物理位置的限制，可以根据功能、部门及应用等因素组织起来，相互之间的通信就好像在同一个网段中一样。

在二层交换机组成的网络中，VLAN 技术实现了本地网络中的数据流量分割，不同的 VLAN 之间不能互相通信，隔离了网络中广播通信的流量，减少了网络中的干扰信息。但如果要实现不同部门的 VLAN 之间的通信，需要借助三层路由技术。在网络发展早期，主要利用路由器的单臂路由技术实现。随着三层交换技术的广泛应用，现在的局域网中更多地借助三层交换技术来实现，通常表现为三层交换机或弱三层交换机产品。

三层交换机在针对第一个 IP 数据报文进行路由处理后，会建立一个 MAC 地址与 IP 地址的映射表；后续具有同样特征的 IP 数据报文再次通过该设备时，将根据此表直接通过二层转发，而不再使用三层路由，从而减少了网络延迟，提高了 IP 数据报文转发的效率，消除了网络传输瓶颈问题。

【实施步骤】

（1）搭建网络环境，配置 IP 地址。

按照图 1-1 搭建某公司的部分办公网络环境。按照不同部门网络地址规划，配置测试计算机 PC1、PC2 的 IP 地址，如图 1-2、图 1-3 所示。

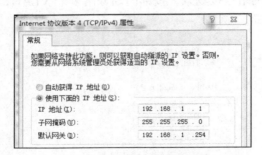

图 1-2　配置 PC1 的 IP 地址

图 1-3　配置 PC2 的 IP 地址

（2）在二层接入交换机 Switch 3 上配置 VLAN 信息。

```
Ruijie>enable
Ruijie#configure terminal
Ruijie(config)#vlan 10
Ruijie(config-vlan)#exit
Ruijie(config)#interface FastEthernet 0/1
Ruijie(config-if-FastEthernet 0/1)#switchport mode trunk
Ruijie(config-if-FastEthernet 0/1)#exit
Ruijie(config)#interface range FastEthernet 0/2-24
Ruijie(config-if-range)#switchport access vlan 10
Ruijie(config-if-range)#exit
```

任务 ❶ 配置跨交换机 VLAN 之间的 SVI 路由

（3）在二层接入交换机 Switch 2 上配置 VLAN 信息。

```
Ruijie>enable
Ruijie#configure terminal
Ruijie(config)#vlan 20
Ruijie(config-vlan)#exit
Ruijie(config)#interface FastEthernet 0/1
Ruijie(config-if-FastEthernet 0/1)#switchport mode trunk
Ruijie(config-if-FastEthernet 0/1)#exit
Ruijie(config)#interface range FastEthernet 0/2-24
Ruijie(config-if-range)#switchport access vlan 20
Ruijie(config-if-range)#exit
```

（4）在交换机 Switch 1 上配置三层 SVI 技术。

```
Ruijie>enable
Ruijie#configure terminal
Ruijie(config)#vlan 10
Ruijie(config-vlan)#exit
Ruijie(config)#interface vlan 10
Ruijie(config-if-vlan 10)#ip address 192.168.1.254 255.255.255.0
Ruijie(config-if-vlan 10)#exit

Ruijie(config)#vlan 20
Ruijie(config-vlan)#exit
Ruijie(config)#interface vlan 20
Ruijie(config-if-vlan 20)#ip address 192.168.2.254 255.255.255.0
Ruijie(config-if-vlan 20)#exit

Ruijie(config)#interface range GigabitEthernet 0/1-2
Ruijie(config-if-range)#switchport mode trunk
Ruijie(config-if-range)#exit
Ruijie(config)#
```

【测试验证】

（1）在 Switch 1 交换机上查看路由表，确认 VLAN 间的路由信息。

```
Ruijie#Show ip route
...
```

（2）在计算机 PC1 上转到 DOS 命令状态，使用 Ping 命令测试两个不同部门网络是否连通，测试网络连通结果如图 1-4 所示。

```
PC1> ping 192.168.2.1
```

测试结果显示，通过在三层交换机上配置 SVI，实现了不同 VLAN 之间的相互通信。

多层交换技术（实践篇）

在网络测试的过程中，需要注意两点：一是如果使用笔记本式计算机来当作测试计算机，就要在测试 Ping 命令的时候将该设备的无线网卡禁用，否则会形成误导数据包传输路径。二是建议关闭计算机上的 Windows 防火墙功能，Windows 防火墙会禁止来自其他网络的 Ping 探测。

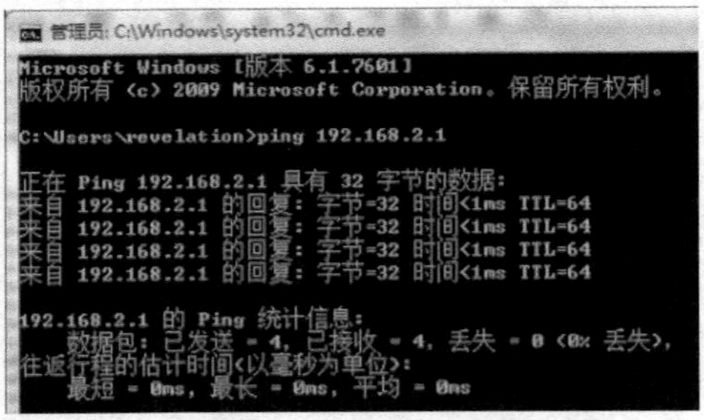

图 1-4　在计算机 PC1 上测试部门网络已连通

【注意事项】

不同 VLAN 之间的 SVI 技术，需要通过三层交换机对数据进行路由转发才可以实现，通过在三层交换机上为各个 VLAN 配置 SVI，利用三层交换机的路由功能实现 VLAN 之间的路由。

最近几年来，为了简化网络配置，实现网络的扁平化管理，出现了能实现 SVI 技术的二层交换机设备，如锐捷网络的 S26 系列交换机、S29 系列交换机。通常把这种交换机称为弱三层交换机，弱三层交换机指具有部分路由功能的二层交换机，可以在弱三层交换机上配置 SVI 或者静态路由，能建立和维护独立的路由表。弱三层交换机最重要的目的是加快大型局域网内部的 IP 数据交换，其所具有的路由功能也是为这一目的服务的，能够做到一次路由，多次转发。在转发 IP 数据报文等规律性的数据流量过程中，可通过硬件芯片实现高速传输。只有路由信息更新、路由表维护、路由计算、路由确定等功能，才由软件实现。

图 1-5 所示的网络拓扑使用锐捷网络的 S29 系列二层交换机产品，实现了两个不同 VLAN 之间的通信，相关配置如下。

```
Ruijie>enable
Ruijie#configure terminal
Ruijie(config)#hostname S2928
S2928(config)#vlan 10
S2928(config-vlan)#vlan 20
S2928(config-vlan)#exit
S2928(config)#interface GigabitEthernet 0/1
S2928(config-if-GigabitEthernet 0/1)#switchport access vlan 10
S2928(config-if-GigabitEthernet 0/1)#exit
S2928(config)#interface GigabitEthernet 0/2
S2928(config-if-GigabitEthernet 0/2)#switchport access vlan 20
```

任务 ❶ 配置跨交换机 VLAN 之间的 SVI 路由

```
S2928(config-if)#exit

S2928(config)#interface vlan 10
S2928(config-if-vlan 10)#ip address 192.168.10.254 255.255.255.0
S2928(config-if)#exit
S2928(config)#interface vlan 20
S2928(config-if-vlan 20)#ip address 192.168.20.254 255.255.255.0
S2928(config-if)#end

S2928#show ip route
...
```

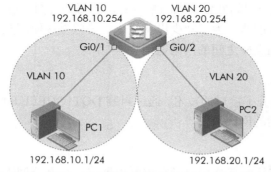

图 1-5　使用 S29 系列二层交换机产品实现两个不同 VLAN 之间的通信

任务 ❷ 单臂路由技术实现不同 VLAN 之间通信

【任务描述】

某电子商务公司是一家有几十台电脑的 SOHO 公司，依托互联网开展业务，组建了互连互通的企业网。为实现公司内部的计算机通过高带宽访问互联网，公司安装了一台专业宽带路由器。为了优化公司的办公接入网络，公司内部办公网通过 VLAN 技术，按照部门划分了多个部门 VLAN 网络。为了减少设备投入，希望直接使用宽带路由器，通过单臂路由技术实现不同部门 VLAN 之间的网络通信。

【任务目标】

掌握在路由器上划分子接口的技术，通过封装 DOT1Q(IEEE 802.1Q)协议，使用单臂路由实现不同 VLAN 之间的网络通信。

【组网拓扑】

图 2-1 所示网络拓扑为某公司通过宽带路由器将公司办公网络接入互联网场景，在路由器的以太网接口上使用单臂路由技术，实现了不同 VLAN 之间的数据通信。

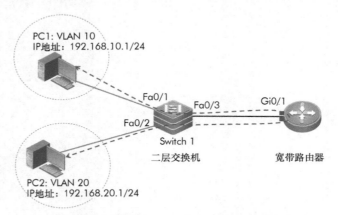

图 2-1 某电子商务公司将办公网络接入互联网场景
（备注：具体的设备连接状况及连接接口信息可根据实际情况决定）

【设备清单】

二层交换机（1 台），路由器（1 台），网线（若干），PC（若干）。

【关键技术】

单臂路由（router-on-a-stick）是指在路由器的一个物理接口上，通过配置子接口（或

任务 ❷ 单臂路由技术实现不同 VLAN 之间通信

逻辑接口）的方式，实现原来相互隔离的不同 VLAN 之间的互连互通。路由器的物理接口可以被划分为多个逻辑接口，这些逻辑接口被形象地称为子接口。值得注意的是，这些逻辑接口不能被单独地开启或关闭。也就是说，当物理接口被开启或关闭时，该接口上的所有子接口也随之被开启或关闭。

在路由器的以太网接口上创建若干个子接口，每个子接口对应一个 VLAN 并作为连接对应 VLAN 的网关。当使用路由器连接到一台划分有多个 VLAN 的二层交换机时，可以通过路由器的一个以太网接口，实现二层交换机上多个 VLAN 之间的互通。

单臂路由技术需要在二层交换机上配置 VLAN，在路由器连接交换机的接口上划分子接口，给相应的 VLAN 设置 IP 地址，实现 VLAN 之间的路由。

【实施步骤】

（1）在二层交换机 Switch 1 上配置不同部门的 VLAN 信息。

```
Ruijie>enable
Ruijie#configure terminal
Ruijie(config)#vlan 10
Ruijie(config-vlan)#vlan 20
Ruijie(config-vlan)#exit
```

（2）在二层交换机 Switch 1 上，将接口划分到相应 VLAN，设置 Trunk 接口。

```
Ruijie(config)#
Ruijie(config)#interface FastEthernet 0/1
Ruijie(config-if-FastEthernet 0/1)#switchport mode access      // 可选配置项
Ruijie(config-if-FastEthernet 0/1)#switchport access vlan 10
Ruijie(config-if-FastEthernet 0/1)#exit
Ruijie(config)#

Ruijie(config)#interface FastEthernet 0/2
Ruijie(config-if-FastEthernet 0/2)#switchport mode access      // 可选配置项
Ruijie(config-if-FastEthernet 0/2)#switchport access vlan 20
Ruijie(config-if-FastEthernet 0/2)#exit
Ruijie(config)#

Ruijie(config)#interface FastEthernet 0/3
Ruijie(config-if-FastEthernet 0/3)#switchport mode trunk
//将接口设置成 Trunk
Ruijie(config-if-FastEthernet 0/3)#exit
Ruijie(config)#
```

（3）在宽带路由器上创建子接口，配置 IP 地址。

```
Ruijie#configure terminal
Ruijie(config)#interface GigabitEthernet 0/0.10      //进入子接口 G0/0.10
Ruijie(config-if-GigabitEthernet 0/0.10)#encapsulation dot1Q 10
```

多层交换技术（实践篇）

```
                                // 指定子接口 G0/0.10 对应 VLAN 10，并配置 Trunk 模式
    Ruijie(config-if-GigabitEthernet 0/0.10)#ip address 192.168.10.254 255.
255.255.0
                                        //配置子接口 G0/0.10 上的 IP 地址
    Ruijie(config-if-GigabitEthernet 0/0.10)#exit
    Ruijie(config)#

    Ruijie(config)#interface GigabitEthernet 0/0.20
    Ruijie(config-if-GigabitEthernet 0/0.20)#encapsulation dot1Q 20
                                // 指定子接口 G0/0.20 对应 VLAN 20，并配置 Trunk 模式
    Ruijie(config-if-GigabitEthernet 0/0.20)#ip address 192.168.20.254 255.
255.255.0
    Ruijie(config-if-GigabitEthernet 0/0.20)#end
    Ruijie(config)#
```

（4）查看交换机的 VLAN 和 Trunk 配置。

① 在接入交换机 Switch 1 上，使用如下命令查看 VLAN 配置信息，如图 2-2 所示。

```
    Ruijie(config)#show vlan
```

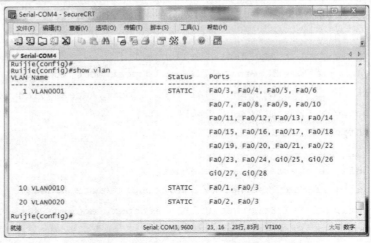

图 2-2 查看 VLAN 配置信息

② 在接入交换机 Switch 1 上，使用如下命令查看 Trunk 接口信息，如图 2-3 所示。

```
    Ruijie(config)#show interfaces FastEthernet 0/3 switchport
```

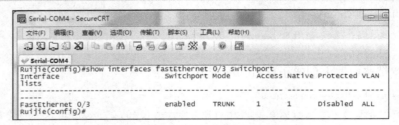

图 2-3 查看 Trunk 接口信息

任务 ❷ 单臂路由技术实现不同 VLAN 之间通信

【测试验证】

按图 2-1 所示的网络拓扑连线并配置测试计算机地址。按照不同子网地址规划，配置测试计算机 PC1 和 PC2 的 IP 地址。

在部门 VLAN 10 网络中的 PC1 上，使用 Ping 命令，测试部门 VLAN 20 网络中 PC2 的 IP 地址和网关地址。测试网络连通结果如图 2-4 所示，不同部门的 VLAN 中设备通过单臂路由实现了连通。

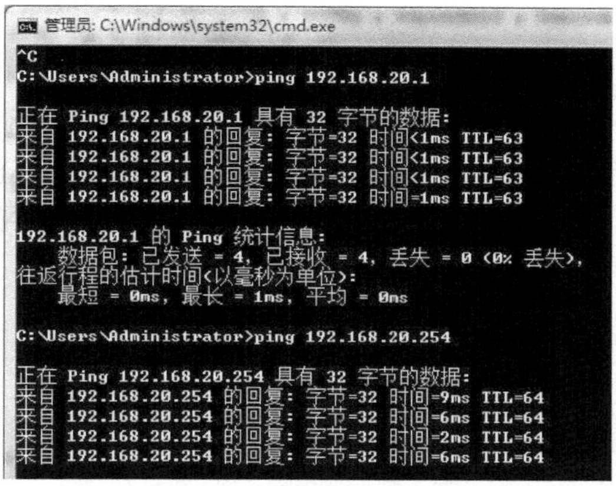

图 2-4 测试网络连通结果

在任务实施的过程中，需要注意两点，一是在给路由器子接口配置 IP 地址之前，一定要先封装 DOT1Q 协议并映射相应的 VLAN；二是各个 VLAN 内的主机，要以相应 VLAN 子接口的 IP 地址作为网关。

【备注】

VLAN 能有效分割局域网，实现各区域网络之间的访问控制。但现实中，往往需要配置某些 VLAN 之间的互连互通。

单臂路由技术是网络发展早期，在三层交换技术没有广泛应用之前，实现不同 VLAN 之间通信的重要技术。

随着三层交换技术的大规模应用，单臂路由技术在有线网络的技术发展中逐渐退出了应用。但在无线局域网技术场景中，当实现不同无线区域网中的用户通过无线 AP 接入到有线网络时，还在广泛应用该技术。

任务 ❸ 修剪骨干链路 VLAN 广播，优化传输效率

【任务描述】

某公司为了加强公司信息化建设，组建了互连互通的企业网络。为了保障公司网络中心骨干交换机上连接的网络服务器的传输效率，希望在骨干链路上实施 VLAN 修剪技术，屏蔽部分和业务无关的部门 VLAN 的信息，优化和改善网络传输，提高网络工作效率。

【任务目标】

了解 VLAN 修剪技术，减少骨干链路上的广播、组播、单播，提升骨干链路上的传输带宽，减少骨干链路上不必要的用户数据信息。

【组网拓扑】

在图 3-1 所示网络拓扑中，两台部门计算机都属于 VLAN 10，可以实现通信。在两台交换机级联的干道口上做 VLAN 修剪，阻止后勤部门 PC1 上的数据通过干道访问网络服务器。

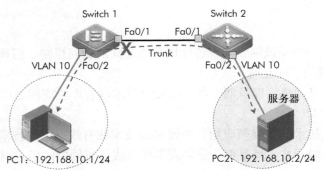

图 3-1　骨干链路修剪拓扑

（备注：具体的设备连接接口信息可根据实际情况决定）

【设备清单】

交换机（两台），网线（若干），PC（若干）。

【关键技术】

为了实现同一个部门 VLAN 跨交换机通信，需要在互连的交换机之间配置干道（Trunk）技术，干道接口允许所有带标签（Tag）的数据帧通过。因此，每一个 VLAN 内部产生的单播、组播和广播信息都会通过干道传输到另一台互连的交换机中。

任务 ❸ 修剪骨干链路 VLAN 广播，优化传输效率

在某些场景中，为了减少干道上不必要的广播信息，优化干道链路上的广播、组播、单播传输，保留带宽，启用了 VLAN 修剪技术，将一些不必要的流量修剪掉，拒绝这些数据流通过干道，减少中继链路上不必要的信息。

VLAN 修剪技术是 Trunk 技术的一项属性，它能减少 Trunk 接口上不必要的信息量。默认情况下，VLAN 修剪功能是关闭的，需要手工打开。

通过在交换机的 Trunk 接口上开启 VLAN 修剪，不仅可提高网络带宽的利用率，还可提高部门网络的安全性，使部门内部含有敏感数据的用户设备与内部网络的其余部门隔离开，从而降低泄露机密信息的可能性。

【实施步骤】

（1）在二层接入交换机 Switch 1 上配置 VLAN 的基本信息。

```
Ruijie#configure terminal
Ruijie(config)#vlan 10
Ruijie(config-vlan)#exit
Ruijie(config)#interface FastEthernet 0/1
Ruijie(config-if-FastEthernet 0/1)#switchport mode trunk
Ruijie(config-if-FastEthernet 0/1)#exit
Ruijie(config)#interface range FastEthernet 0/2
Ruijie(config-if-FastEthernet 0/2)#switchport access vlan 10
Ruijie(config-if-FastEthernet 0/2)#end
```

（2）在汇聚层交换机 Switch 2 上配置 VLAN 的基本信息。

```
Ruijie#configure terminal
Ruijie(config)#vlan 10
Ruijie(config-vlan)#exit
Ruijie(config)#interface FastEthernet 0/1
Ruijie(config-if-FastEthernet 0/1)#switchport mode trunk
Ruijie(config-if-FastEthernet 0/1)#exit
Ruijie(config)#interface FastEthernet 0/2
Ruijie(config-if-FastEthernet 0/2)#switchport access vlan 10
Ruijie(config-if-FastEthernet 0/2)#end
```

（3）测试连通性。分别给两台测试计算机配置同一子网的 IP 地址，使用 Ping 命令测试网络连通性，结果如图 3-2 所示，说明网络连通正常。

（4）在接入交换机 Switch 1 上配置干道接口 VLAN 修剪。

```
Ruijie(config)#interface FastEthernet 0/1
Ruijie(config-if-FastEthernet 0/1)#switchport mode trunk
Ruijie(config-if-FastEthernet 0/1)#switchport trunk allowed vlan remove 10
             // 不允许后勤部门 VLAN 10 的数据通过，需要在互连的干道接口上配置
Ruijie(config-if-FastEthernet 0/1)#exit
```

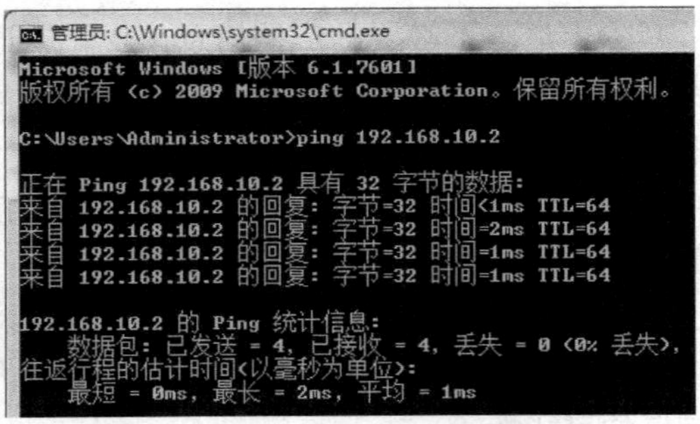

图 3-2 测试网络连通性的结果

（5）在汇聚交换机 Switch 2 上配置干道接口 VLAN 修剪。

```
Ruijie(config)#interface FastEthernet 0/1
Ruijie(config-if-FastEthernet 0/1)#switchport mode trunk
Ruijie(config-if-FastEthernet 0/1)#switchport trunk allowed vlan remove 10
            // 不允许后勤部门 VLAN 10 的数据通过，需要在互连的干道接口上配置
Ruijie(config-if-FastEthernet 0/1)#exit
```

（6）干道修剪屏蔽 VLAN 连通。

在接入交换机和汇聚交换机的干道接口上实施 VLAN 修剪后，继续使用后勤部门测试计算机 PC1，测试网络中心服务器 PC2。

使用 Ping 命令测试网络连通性，结果如图 3-3 所示，说明来自后勤部门 VLAN 10 中的数据已经无法通过干道链路，后勤部门的 PC1 也无法 Ping 通。

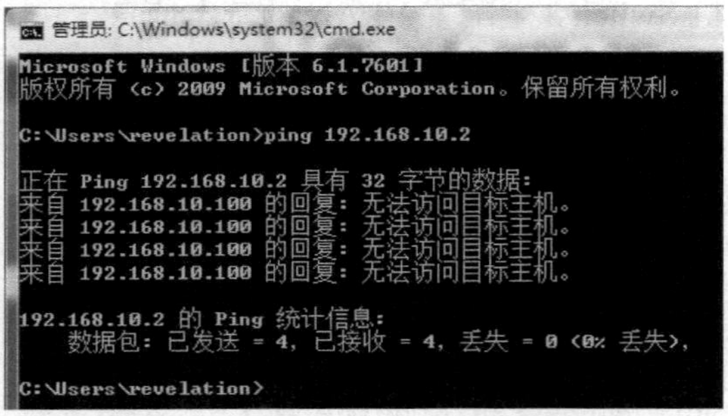

图 3-3 干道链路修剪屏蔽 VLAN 网络后的结果

任务 ④ 配置 PVLAN，实现 VLAN 内部隔离

【任务描述】

某大学期末考试需要机考，为了避免考试中相邻机位之间通过本地网络传输考试内容，教务处希望在考试期间实施考生机位之间的网络隔离，以防止作弊，即同一机房中相邻机位之间不能通信，但需要每一位学生都能把考试结果提交到同一机房中的教师机上。

按机位划分 VLAN，可以实现相邻机位之间的隔离，但这样一间 50 台机位的机房需要划分 50 个 VLAN，不仅麻烦，还占用了学院很多 VLAN 资源。实施私有 VLAN（Private VLAN，PVLAN）技术，经过简单配置就可以实现相同 VLAN 内部数据的隔离。

【任务目标】

了解 VLAN 的特殊应用，掌握 PVLAN 技术原理及应用场景，通过配置 PVLAN 技术功能实现 VLAN 内部的数据隔离。

【组网拓扑】

图 4-1 所示网络拓扑为某学院某间多媒体机房网络场景，使用了两台交换机与 4 台测试计算机，配置了主 VLAN 10、团体 VLAN 20、隔离 VLAN 30，实现 PVLAN 安全隔离功能，实施同一个 VLAN 内部机位之间的安全隔离。

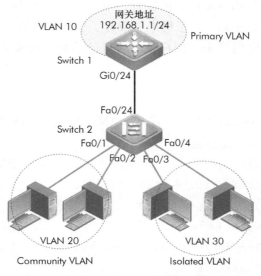

图 4-1　某学院某间多媒体机房网络场景

（备注：具体的设备连接接口信息可根据实际情况决定）

多层交换技术（实践篇）

【设备清单】

三层交换机（1台），二层交换机（1台），网线（若干），PC（若干）。

【关键技术】

PVLAN 采用两层 VLAN 隔离技术实现 VLAN 中的 VLAN 效果，只有上层 VLAN 全局可见，下层 VLAN 相互隔离。其中，每个 PVLAN 包含两种类型的 VLAN，分别是 Primary VLAN（主 VLAN）和 Secondary VLAN（辅助 VLAN）。辅助 VLAN 又包含两种类型，即 Isolated VLAN（隔离 VLAN）和 Community VLAN（团体 VLAN）。

在 PVLAN 技术中，连接的交换机端口也有两种类型：一般把与用户相连的、分配到辅助 VLAN 中的端口称为隔离端口（Isolated Port）和团队端口（Community Port）；把与上级交换机或网关设备相连的、分配到主 VLAN 中的端口称为混合端口（Promiscuous Port）。这些端口对应不同的 VLAN 类型，其中，辅助 VLAN 中的 Isolated Port 属于辅助 VLAN 中的 Isolated PVLAN；Community Port 端口属于辅助 VLAN 中的 Community PVLAN。这两类辅助 VLAN 端口都需要和主 VLAN 中的 Promiscuous Port 通信。

PVLAN 技术规定：在 Isolated PVLAN 中，Isolated Port 之间不能交换数据，只能与 Promiscuous Port 进行通信；在 Community PVLAN 中，Community Port 不仅可以和 Promiscuous Port 通信，彼此之间也可以通信。

PVLAN 将同一个 VLAN 内部的端口分为两类：与用户相连的端口称为隔离端口，每个隔离端口可划分到不同的 PVLAN；上行与上级交换机或网关设备相连的端口为混合端口，混合端口划分到 Primary VLAN。

隔离端口只能与混合端口通信，相互之间不能通信。一个 Primary VLAN 可以包含多个 PVLAN 中包含的端口和上行端口，这样对于上层交换机来说，可以认为下层交换机中只有几个 Primary VLAN，而不必关心 Primary VLAN 中的端口实际所属的 VLAN。

PVLAN 技术简化了 VLAN 内部的配置，节省了 VLAN 资源。一个 Primary VLAN 中包含的所有 PVLAN 处于同一个子网中，节省了子网数目和 IP 地址资源，这样可将同一个 VLAN 中的端口隔离开来，用户只能与自己的默认网关通信，网络的安全性得到了保障。

【实施步骤】

（1）在二层交换机 Switch 2 上，配置 PVLAN 中相关 VLAN 关联。

```
Ruijie#configure terminal
Ruijie(config)#vlan 20                              // 创建辅助 VLAN 20
Ruijie(config-vlan)#private-vlan community          // VLAN 20 属性为团体 VLAN
Ruijie(config-vlan)#exit
Ruijie(config)#vlan 30                              // 创建辅助 VLAN 30
Ruijie(config-vlan)#private-vlan isolated           // VLAN 30 属性为隔离 VLAN
Ruijie(config-vlan)#exit
Ruijie(config)#vlan 10                              // 创建主 VLAN 10
```

任务 ❹ 配置 PVLAN，实现 VLAN 内部隔离

```
Ruijie(config-vlan)#private-vlan primary        // VLAN 10 属性为主 VLAN
Ruijie(config-vlan)#private-vlan association 20, 30
                                               // 主 VLAN 10 关联 Secondary VLAN
Ruijie(config-vlan)#exit
```

（2）在二层交换机 Switch 2 上，配置 PVLAN 各个端口的属性。

```
Ruijie#configure terminal
Ruijie(config)#interface range FastEthernet 0/1-2
Ruijie(config-if-range)#switchport mode private-vlan host
                       // 配置为私有 VLAN 端口
Ruijie(config-if-range)#switchport private-vlan host-association 10 20
                       // 将端口加入团体 VLAN 20 并和主 VLAN 10 关联
Ruijie(config-if-range)#exit

Ruijie(config)#interface range FastEthernet 0/3-4
Ruijie(config-if-range)#switchport mode private-vlan host
                       // 配置为私有 VLAN 端口
Ruijie(config-if-range)#switchport private-vlan host-association 10 30
                       // 将端口加入团体 VLAN 30 并和主 VLAN 10 关联
Ruijie(config-if-range)#exit

Ruijie(config)#interface FastEthernet 0/24
Ruijie(config-if-FastEthernet 0/24)#switchport mode trunk
Ruijie(config-if-FastEthernet 0/24)#switchport mode private-vlan promiscuous
// 配置上连端口为混合端口，指定主、辅 VLAN 的映射关系，使私有 VLAN 报文通过端口发送
```

（3）在三层汇聚交换机 Switch 1 上，配置 VLAN 基本属性。

```
Ruijie(config)#vlan 10
Ruijie(config-vlan)#exit
Ruijie(config)#interface vlan 10       // 配置主 VLAN 地址，作为默认网关
Ruijie(config-if-vlan 10)#ip address 192.168.1.1 255.255.255.0
Ruijie(config-if-vlan 10)#exit

Ruijie(config)#interface GigabitEthernet0/24
Ruijie(config-if-GigabitEthernet0/24)#switchport mode trunk
Ruijie(config-if-GigabitEthernet0/24)#exit
```

备注：PVLAN 需要在三层交换机上进行配置，在部分弱三层交换机上也可以实施。

【测试验证】

（1）配置测试计算机 IP 地址。

给所有计算机配置 IP 地址。按照 PVLAN 的特性，所有计算机使用同一网段地址。其

中，PC1 的 IP 地址为 192.168.1.2/24；PC2 的 IP 地址为 192.168.1.3/24；PC3 的 IP 地址为 192.168.1.4/24；PC4 的 IP 地址为 192.168.1.5/24，默认网关都是 192.168.1.1。

（2）测试团体 VLAN 内部计算机之间能否通信。

在团体 VLAN 中，打开 PC1，转到 DOS 状态，使用 Ping 命令测试其与 PC2 的网络连通状况，按照 PVLAN 中关于团体 VLAN 的属性规定，即团体 VLAN 内部设备实现通信，结果如图 4-2 所示。

（3）测试独立 VLAN 和团体 VLAN 能否通信。

在团队 VLAN 中，打开 PC1，转到 DOS 状态，使用 Ping 命令测试其与 PC3 的网络连通状况，测得 PC1 和 PC3 之间互相隔离，如图 4-3 所示。

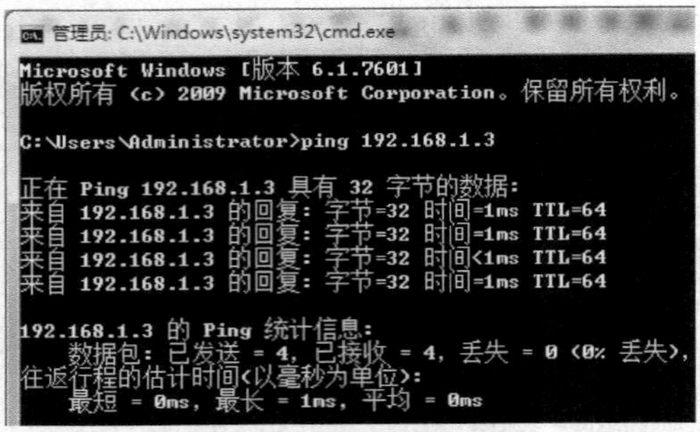

图 4-2 团体 VLAN 内部设备可实现通信　　图 4-3 独立 VLAN 和团体 VLAN 互相隔离

（4）测试团体 VLAN 和网关能否通信。

用同样的方式，测试 PC1 同网关的通信，测得团体 VLAN 和网关之间可以通信，如图 4-4 所示。

（5）测试独立 VLAN 内部计算机之间能否通信。

在独立 VLAN 中，打开 PC3，转到 DOS 状态，使用 Ping 命令测试其与 PC4 的网络连通状况，测得独立 VLAN 内部计算机之间互相隔离，如图 4-5 所示。

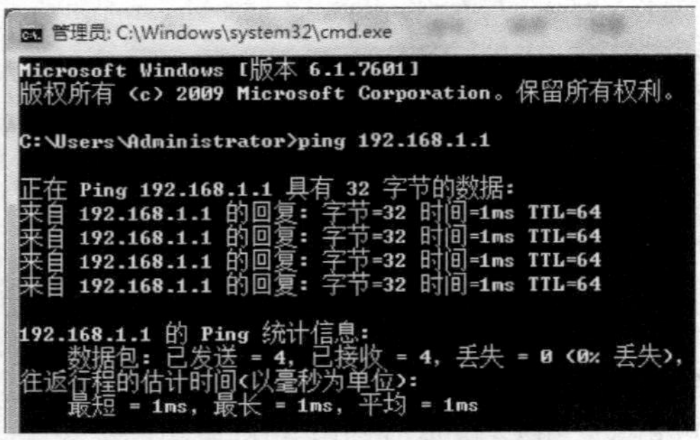

图 4-4 团体 VLAN 和网关可实现通信 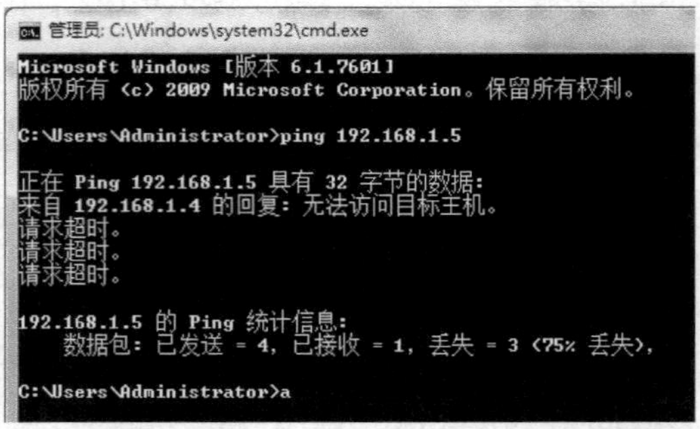　　图 4-5 独立 VLAN 内部计算机之间不能通信

在 PC3 中使用 Ping 命令测试网关，测得 Isolated 端口同 Promiscuous 端口之间可以通信，如图 4-6 所示。

需要注意的是，部分交换机产品不具有 PVLAN 性能，需要根据具体产品性能完成上述任务。

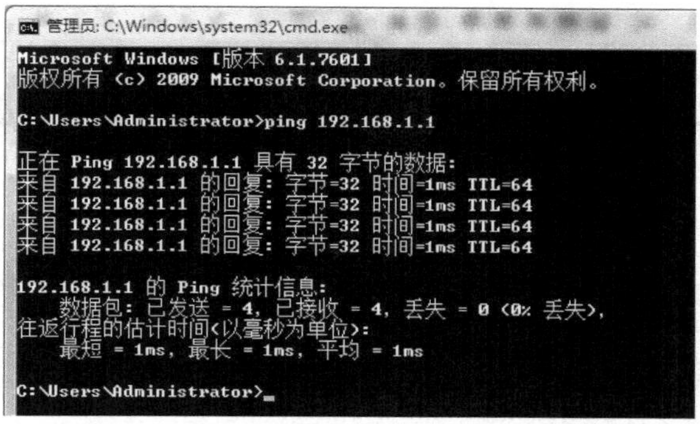

任务 ❹ 配置 PVLAN，实现 VLAN 内部隔离

图 4-6 Isolated 端口同 Promiscuous 端口之间可以通信

任务 5 使用 Super VLAN 技术优化 IP 地址

【任务描述】

某公司在全国各地区都建有分公司，各分公司通过 VPN 虚拟专网技术和北京的总公司网络连通。其中，上海分公司出于安全需要，按部门划分有多个部门 VLAN，分公司网络按部门安全隔离。出于 IP 地址的规划需要，上海分公司使用总公司统一分配的一个子网 IP 地址段，只能通过 Super VLAN 技术使所有部门 VLAN 共享一个独立子网，实现网络互连互通。

【任务目标】

了解 Super VLAN 技术，通过配置 Super VLAN 技术掌握 Super VLAN 应用场景。

【组网拓扑】

图 5-1 所示网络拓扑为某分公司使用 Super VLAN 技术部署的场景。其中，三层交换机 Switch 1 作为用户网关设备，通过 Trunk 接口下连二层交换机 Switch 2、Switch 3。二层交换机按照部门划分 VLAN 实现了二层隔离，所有 VLAN 共享一个 IP 网关，通过三层交换设备与外网通信。

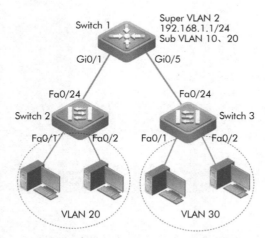

图 5-1 某分公司使用 Super VLAN 技术部署的场景
（备注：实际设备接口可根据现场情况选择，并做相应修改）

【设备清单】

二层交换机（两台），三层交换机（1 台），网线（若干），PC（若干）。

任务 ❺ 使用 Super VLAN 技术优化 IP 地址

【关键技术】

传统的网络规划中，要为每个 VLAN 分配一个 IP 子网。每分配一个 IP 子网就有 3 个 IP 地址被占用，分别作为子网的网络号、广播地址和默认网关。即使有些子网中有大量未分配的 IP 地址，也无法给其他子网中的用户使用，形成 IP 地址的浪费。Super VLAN 有效地解决了这个问题，它把多个 VLAN（也称为子 VLAN，Sub VLAN）聚合成一个 Super VLAN，这些 VLAN 都使用同一个子网的 IP 地址和默认网关。

利用 Super VLAN 技术，只需为 Super VLAN 分配一个 IP 子网，所有子 VLAN 就可以灵活分配 Super VLAN 子网中的 IP 地址，使用 Super VLAN 的默认网关。同时，每个子 VLAN 还是一个独立的广播域，仍能实现不同 VLAN 之间的隔离，子 VLAN 之间的通信需要通过 Super VLAN 上三层路由实现。

由于各个子 VLAN 不需要规划独立的子网网段，因此 IP 利用率得到提高，IP 地址使用效率得到优化。

【实施步骤】

（1）在三层交换机 Switch 1 上分别创建 VLAN 2、VLAN 10、VLAN 20。

```
Ruijie#configure terminal
Ruijie(config)#vlan 2              // 创建 Super VLAN 2
Ruijie(config-vlan)#exit
Ruijie(config)#vlan 10             // 创建 Sub VLAN 10
Ruijie(config-vlan)#exit
Ruijie(config)#vlan 20             // 创建 Sub VLAN 20
Ruijie(config-vlan)#exit
```

（2）在三层交换机 Switch 1 上设置 VLAN 2 为 Super VLAN，对应 Sub VLAN 为 VLAN 10、VLAN 20。

```
Ruijie(config)#vlan 2
Ruijie(config-vlan)#supervlan      // 设置 VLAN 2 为 Super VLAN
Ruijie(config-vlan)#subvlan 10,20
                     // 设置 Super VLAN 对应的 Sub VLAN 为 VLAN 10、VLAN 20
Ruijie(config-vlan)#exit
```

（3）在三层交换机 Switch 1 上配置三层 SVI 地址。

```
Ruijie(config)#interface vlan 2
Ruijie(config-if-vlan 2)#ip address 192.168.196.1 255.255.255.0
Ruijie(config-if-vlan 2)#no shutdown
Ruijie(config-if-vlan 2)#exit
Ruijie(config)#
```

（4）在三层交换机 Switch 1 上配置分配的 IP 地址范围。

设置 VLAN 10 的 IP 地址范围为 192.168.196.10 /24～192.168.196.50/ 24，设置 VLAN 20 的 IP 地址范围为 192.168.196.60/24～192.168.196.100 /24。

```
Ruijie(config)#vlan 10
Ruijie(config-vlan)#subvlan-address-range 192.168.196.10    192.168.196.50
       // 分配 VLAN 10 的 IP 地址范围为 192.168.196.10/24 ~192.168.196.50/24
Ruijie(config-vlan)#exit
Ruijie(config)#vlan 20
Ruijie(config-vlan)#subvlan-address-range 192.168.196.60    192.168.196.100
       // 分配 VLAN 20 的 IP 地址范围为 192.168.196.60/24 ~192.168.196.100/24
Ruijie(config-vlan)#exit
```

（5）在三层交换机 Switch 1 上设置连接交换机 Switch 2、Switch 3 的接口为 Trunk 接口。

```
Ruijie(config)#interface GigabitEthernet 0/1
Ruijie(config-if-GigabitEthernet 0/1)#switchport mode trunk
Ruijie(config-if-GigabitEthernet 0/1)#exit
Ruijie(config)#interface GigabitEthernet 0/5
Ruijie(config-if-GigabitEthernet 0/5)#switchport mode trunk
Ruijie(config-if-GigabitEthernet 0/5)#exit
```

（6）在二层接入交换机 Switch 2 上配置接口和 VLAN 信息。

```
Ruijie(config)#vlan 10
Ruijie(config-vlan)#exit
Ruijie(config)#interface FastEthernet 0/1    // 分配连接用户接口到 VLAN 10
Ruijie(config-if-FastEthernet 0/1)#switchport access vlan 10
Ruijie(config-if-FastEthernet 0/1)#exit
Ruijie(config)#interface FastEthernet 0/2    // 分配连接用户接口到 VLAN 10
Ruijie(config-if-FastEthernet 0/2)#switchport access vlan 10
Ruijie(config-if-FastEthernet 0/2)#exit
Ruijie(config)#interface FastEthernet 0/24    // 配置上连接口为 Trunk 接口
Ruijie(config-if-FastEthernet 0/24)#switchport mode trunk
Ruijie(config-if-FastEthernet 0/24)#exit
```

（7）在二层接入交换机 Switch 3 上配置接口和 VLAN 信息。

```
Ruijie(config)#vlan 20
Ruijie(config-vlan)#exit
Ruijie(config)#interface FastEthernet 0/1    // 分配连接用户接口到 VLAN 20
Ruijie(config-if-FastEthernet 0/1)#switchport access vlan 20
Ruijie(config-if-FastEthernet 0/1)#exit
Ruijie(config)#interface FastEthernet 0/2    // 分配连接用户接口到 VLAN 20
Ruijie(config-if-FastEthernet 0/2)#switchport access vlan 20
Ruijie(config-if-FastEthernet 0/2)#exit
Ruijie(config)#interface FastEthernet 0/24    // 配置上连接口为 Trunk 接口
Ruijie(config-if-FastEthernet 0/24)#switchport mode trunk
```

任务 ❺ 使用 Super VLAN 技术优化 IP 地址

```
Ruijie(config-if-FastEthernet 0/24)#exit
```
（8）配置 Switch 1 交换机的 ARP 代理功能（可选）。

默认情况下，具有 Super VLAN 功能的交换机的 ARP 代理功能是开启的，这样下面连接的 Sub VLAN 之间的设备可以通过 ARP 代理实现三层互访。如果需要阻止 Sub VLAN 之间的互访，则应关闭 Super VLAN 的 ARP 代理功能。

```
Ruijie(config)#vlan 2                   //进入 Super VLAN 配置模式
Ruijie(config-vlan)#no proxy-arp        //关闭 Super VLAN 的 ARP 代理功能
```

【测试验证】

（1）在三层交换机 Switch 1 上查看 Super VLAN 配置信息，如图 5-2 所示。

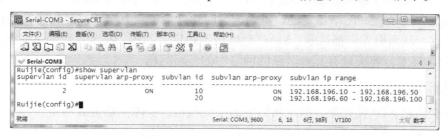

图 5-2 Super VLAN 配置信息

（2）按照 VLAN 10 中的 IP 地址规划 192.168.196.10 /24～192.168.196.50/ 24，为 VLAN 10 中的测试主机配置 IP 地址信息。其中，网关地址统一为 192.168.196.1/24。

（3）按照 VLAN 20 中的 IP 地址规划 192.168.196.60/24～192.168.196.100 /24，为 VLAN 20 中的测试主机配置 IP 地址信息。其中，网关地址统一为 192.168.196.1/24。

（4）在 VLAN 10 中的测试主机上使用"Ping 192.168.196.61"命令，测试与 VLAN 20 中测试主机之间的连通状况。默认情况下，三层交换机 Switch 1 上的 ARP 代理功能开启，所以能 Ping 通。不同的 Sub VLAN 之间通过 Super VLAN 实现通信，如图 5-3 所示。

（5）如上配置，在三层交换机 Switch 1 上关闭 ARP 代理功能。继续在 VLAN 10 的测试主机上使用"Ping 192.168.196.61"命令，测试与 VLAN 20 中测试主机之间的连通状况，关闭 ARP 代理后，不同的 Sub VLAN 之间无法通信，如图 5-3 所示。

图 5-3 不同 Sub VLAN 之间的连通性测试

任务 ❻ 使用 STP 技术保障网络健壮稳定

【任务描述】

某网络中心为提高网络稳定性，使用多台交换机互连构成环形拓扑，实现核心网络的冗余和备份。为了提高网络稳定性，网络构建时通常会提供冗余链路，但是冗余链路会形成物理环路，从而引发广播风暴、MAC 地址表不稳定等问题，甚至导致网络瘫痪。在网络中运行 STP 技术，解决具有冗余链路的网络中的环路问题。

【任务目标】

配置交换机 STP 优先级，指定网络中的根交换机，保障网络健壮稳定。

【组网拓扑】

图 6-1 所示网络拓扑为某网络中心多台交换机实现冗余和备份的场景，启用 STP 技术以增强网络的稳健性。

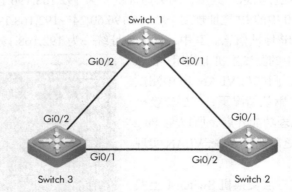

图 6-1 某网络中心多台交换机实现冗余和备份的场景
（备注：具体的设备连接接口信息可根据实际情况决定）

【设备清单】

交换机（3 台），网线（若干），PC（若干）。

【关键技术】

STP 技术在网络中先采用生成树算法，通过交换机优先级等信息选举出一台根交换机，再以根交换机为根节点在网络中形成一棵没有环路的树，从而解决环路引发的问题。当网络中的主要链路出现故障时，STP 技术会重新在网络进行生成树计算，将备份链路恢复，从而达到链路冗余的效果。

任务 ❻ 使用 STP 技术保障网络健壮稳定

【实施步骤】

(1) 分别在三台交换机上启用生成树协议，设置模式为 STP。

```
Ruijie(config)#
Ruijie(config)#hostname Switch 1
Switch 1(config)#spanning-tree
Switch 1(config)#spanning-tree mode stp

Ruijie(config)#
Ruijie(config)#hostname Switch 2
Switch 2(config)#spanning-tree
Switch 2(config)#spanning-tree mode stp

Ruijie(config)#
Ruijie(config)#hostname Switch 3
Switch 3(config)#spanning-tree
Switch 3(config)#spanning-tree mode stp
```

(2) 指定交换机 Switch 1 为根交换机，配置优先级。

```
Switch 1(config)#spanning-tree priority 4096
                    //指定交换机 Switch 1 的优先级为 4096
```

(3) 在交换机 Switch 1 上查看生成树选举结果。

```
Switch 1#show spanning-tree            //查看交换机上的 spanning-tree 配置
StpVersion : STP
SysStpStatus : ENABLED                 //表示交换机的 STP 技术已启用
MaxAge : 20
HelloTime : 2
ForwardDelay : 15
BridgeMaxAge : 20
BridgeHelloTime : 2
BridgeForwardDelay : 15
MaxHops: 20
TxHoldCount : 3
PathCostMethod : Long
BPDUGuard : Disabled
BPDUFilter : Disabled
BridgeAddr : 00d0.f882.f4a1
Priority: 4096                         //显示交换机优先级为 4096
TimeSinceTopologyChange : 0d:0h:1m:12s
TopologyChanges : 2
DesignatedRoot : 1000.00d0.f882.f4a1
RootCost : 0                           //显示交换机到根交换机的开销
```

```
RootPort : 0                                    //显示根端口, 0 表示无根端口
// 从上述 show 命令输出结果可以看出, 交换机 Switch 1 为根交换机, 没有根端口, 根开销为 0

Switch 2#show spanning-tree
StpVersion : STP
SysStpStatus : ENABLED
MaxAge : 20
HelloTime : 2
ForwardDelay : 15
BridgeMaxAge : 20
BridgeHelloTime : 2
BridgeForwardDelay : 15
MaxHops: 20
TxHoldCount : 3
PathCostMethod : Long
BPDUGuard : Disabled
BPDUFilter : Disabled
BridgeAddr : 00d0.f821.a542
Priority: 32768
TimeSinceTopologyChange : 0d:0h:1m:32s
TopologyChanges : 1
DesignatedRoot : 1000.00d0.f882.f4a1
RootCost : 200000                               //显示到达根交换机的开销是 200000
RootPort : 1                                    //显示根端口是编号为 1 的端口
// 从上述 show 命令输出结果可以看出, 交换机 Switch 2 为非根交换机

Switch 3#show spanning-tree
StpVersion : STP
SysStpStatus : ENABLED
MaxAge : 20
HelloTime : 2
ForwardDelay : 15
BridgeMaxAge : 20
BridgeHelloTime : 2
BridgeForwardDelay : 15
MaxHops: 20
TxHoldCount : 3
PathCostMethod : Long
BPDUGuard : Disabled
BPDUFilter : Disabled
BridgeAddr : 00d0.f8b4.e54b
```

任务 ❻ 使用 STP 技术保障网络健壮稳定

```
Priority: 32768
TimeSinceTopologyChange : 0d:0h:14m:48s
TopologyChanges : 0
DesignatedRoot : 1000.00d0.f882.f4a1
RootCost : 200000                        // 显示到达根交换机的开销是 200000
RootPort : 2                              // 显示根端口是 2 号端口
/* 从显示的结果可以看到，Switch 1 没有根端口，在生成树网络中只有根交换机是没有根端口
的，因此从结果中可以判断出 Switch 1 为根交换机，其他交换机为非根交换机 */
```

【注意事项】

（1）使用过程中，需要先在交换机上启用生成树协议，再连接拓扑，否则可能会引发环路。

（2）配置交换机优先级时需要注意，优先级取值是 0～61440，且为 0 或者 4096 的整数倍。

任务 7 使用 RSTP 技术加快网络收敛

【任务描述】

某学院的网络中心使用冗余链路增加备份，通过 STP 避免骨干链路的环路。但在运营中发现当网络拓扑发生变化时，骨干网络的收敛很慢，在网络完成收敛之前，网络中断时间很长。配置 STP 协议的网络收敛需要 30~50s 的时间，在很多大型网络中，这个时间是难以忍受的，RSTP 可以很好地解决这个问题，将网络中断后的收敛时间缩短到 1s 以内。

【任务目标】

理解快速生成树协议的配置及原理。

【组网拓扑】

图 7-1 所示为某学院网络中心骨干拓扑，使用 RSTP 技术实现网络快速收敛。

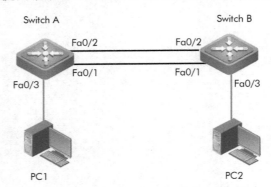

图 7-1 某学院网络中心骨干拓扑

（备注：具体的设备连接接口信息可根据具体情况决定）

【设备清单】

交换机（两台），网线（若干），PC（若干）。

【关键技术】

在交换网络中需要提供冗余备份链路，早期使用 STP 协议解决交换网络中的环路问题。STP 利用生成树算法，在存在交换环路的网络中生成一个没有环路的树形拓扑。STP 算法将交换网络中冗余的链路在逻辑上断开，当主要链路出现故障时，能够自动切换到备份链路，保证数据正常转发。

任务 ❼ 使用 RSTP 技术加快网络收敛

STP 的缺点是收敛时间长，从主要链路出现故障，到切换至备份链路，需要 50s 的时间。

RSTP 在 STP 的基础上增加了替换端口（Alternate Port）和备份端口（Backup Port）两种端口角色，分别作为根端口（Root Port）和指定端口（Designated Port）的备份端口。当根端口或指定端口出现故障时，备份端口不需要经过 50s 的收敛时间就可以直接切换到替换端口或备份端口，从而实现小于 1s 的快速收敛。

【实施步骤】

（1）在交换机 Switch A 上完成 VLAN 划分及 Trunk 配置。

```
Ruijie#configure terminal
Ruijie(config)#hostname Switch A
Switch A(config)#vlan 10
Switch A(config-vlan)#exit

Switch A(config)#interface GigabitEthernet0/3     // 划分计算机到 VLAN 中
Switch A(config-if-GigabitEthernet0/3)#switchport access vlan 10
Switch A(config-if-GigabitEthernet0/3)#exit

Switch A(config)#interface range GigabitEthernet0/1-2
Switch A(config-if-range)#switchport mode trunk     // 设置干道端口
Switch A(config-if-range)#exit
```

（2）在交换机 Switch A 上完成交换机 Switch A 的 RSTP 配置。

```
Switch A(config)#spanning-tree
Switch A(config)#spanning-tree mode rstp     // 指定生成树协议的类型为 RSTP
Switch A(config)#exit
```

（3）在交换机 Switch B 上完成 VLAN 划分及 Trunk 配置。

```
Ruijie#configure terminal
Ruijie(config)#hostname Switch B
Switch B(config)#vlan 10
Switch B(config-vlan)#exit

Switch B(config)#interface GigabitEthernet0/3     // 划分计算机到 VLAN 中
Switch B(config-if-GigabitEthernet0/3)#switchport access vlan 10
Switch B(config-if-GigabitEthernet0/3)#exit
Switch B(config)#interface range GigabitEthernet 0/1-2
Switch B(config-if-range)#switchport mode trunk     // 设置干道端口
Switch B(config-if-range)#exit
```

（4）在交换机 Switch B 上完成交换机 Switch B 的 RSTP 配置。

```
Switch B(config)#
Switch B(config)#spanning-tree
```

```
Switch B(config)#spanning-tree mode rstp        // 指定生成树协议的类型为RSTP
Switch B(config)#exit
```

（5）设置交换机Switch A的优先级，指定Switch A为根交换机。

```
Switch A(config)#
Switch A(config)#spanning-tree priority 4096
                         // 设置交换机Switch A的优先级为4096，使其成为根交换机
```

（6）查看交换机Switch A的端口和RSTP状态。

```
Switch A#show spanning-tree
StpVersion : RSTP                    // 生成树协议的版本
SysStpStatus : Enabled               // 生成树协议的运行状态，Enabled表示开启状态
BaseNumPorts : 24
MaxAge : 20
HelloTime : 2
ForwardDelay : 15
BridgeMaxAge : 20
BridgeHelloTime : 2
BridgeForwardDelay : 15
MaxHops : 20
TxHoldCount : 3
PathCostMethod : Long
BPDUGuard : Disabled
BPDUFilter : Disabled
BridgeAddr : 00d0.f8ef.9e89
Priority : 4096                      // 显示交换机的优先级
TimeSinceTopologyChange : 0d:0h:13m:43s
TopologyChanges : 0
DesignatedRoot : 200000D0F8EF9E89
RootCost : 0
RootPort : 0
// 从show命令的输出结果可以看到，交换机Switch A为根交换机
```

（7）查看交换机Switch B的端口和RSTP状态。

```
Switch B#show spanning-tree
StpVersion : RSTP                    // 生成树协议的版本
SysStpStatus : Enabled               // 生成树协议的运行状态，Enabled表示开启状态
BaseNumPorts : 24
MaxAge : 20
HelloTime : 2
ForwardDelay : 15
BridgeMaxAge : 20
BridgeHelloTime : 2
```

任务 ❼ 使用 RSTP 技术加快网络收敛

```
BridgeForwardDelay : 15
MaxHops : 20
TxHoldCount : 3
PathCostMethod : Long
BPDUGuard : Disabled
BPDUFilter : Disabled
BridgeAddr : 00d0.f8e0.9c81
Priority : 32768                       // 显示交换机的优先级
TimeSinceTopologyChange : 0d:0h:11m:39s
TopologyChanges : 0
DesignatedRoot : 100000D0F8EF9E89
RootCost : 200000                      // 交换机到达根交换机的开销
RootPort : Fa0/1
// 从 show 命令的输出结果可以看到，交换机 Switch B 为非根交换机，根端口为 Fa0/1
```

（8）查看交换机 Switch B 端口 1 和端口 2 的状态。

```
Switch B#show spanning-tree interface FastEthernet 0/1
PortAdminPortfast : Disabled
PortOperPortfast : Disabled
PortAdminLinkType : auto
PortOperLinkType : point-to-point
PortBPDUGuard: Disabled
PortBPDUFilter: Disabled
PortState : forwarding         // 交换机 Switch B 的端口 Fa0/1 处于转发状态
PortPriority : 128
PortDesignatedRoot : 200000D0F8EF9E89
PortDesignatedCost : 0
PortDesignatedBridge : 200000D0F8EF9E89
PortDesignatedPort : 8001
PortForwardTransitions : 3
PortAdminPathCost : 0
PortOperPathCost : 200000
PortRole : rootPort            // 显示端口角色为根端口
/* 从上述 show 命令输出结果可以看到，交换机 Switch B 的端口 Fa0/1 角色为根端口，处于
转发状态 */

Switch B#show spanning-tree interface FastEthernet 0/2
                               // 显示交换机 Switch B 的端口 Fa0/2 的状态
PortAdminPortfast : Disabled
```

多层交换技术（实践篇）

```
PortOperPortfast : Disabled
PortAdminLinkType : auto
PortOperLinkType : point-to-point
PortBPDUGuard: Disabled
PortBPDUFilter: Disabled
PortState : discarding          // 交换机 Switch B 的端口 Fa0/2 处于阻塞状态
PortPriority : 128
PortDesignatedRoot : 200000D0F8EF9E89
PortDesignatedCost : 200000
PortDesignatedBridge : 800000D0F8EF9D09
PortDesignatedPort : 8002
PortForwardTransitions : 3
PortAdminPathCost : 0
PortOperPathCost : 200000
PortRole : alternatePort        // 交换机 Switch B 的 Fa0/2 端口为根端口的替换端口
/* 从上述 show 命令的输出结果可以看到，交换机 Switch B 的端口 Fa0/2 角色为替换端口，处于
阻塞状态 */
```

【测试验证】

把 Switch A 与 Switch B 之间的一条链路 DOWN 掉（如拔掉网线，或者打开该端口，使用"shutdown"命令），验证连接在交换机上的 PC1 与 PC2 之间能否互相 Ping 通，并观察丢包情况。

（1）如图 7-2 所示，在 PC1 上使用 Ping 命令，测试连接 PC2，测试结果为能 Ping 通（其中，PC1 的 IP 地址为 192.168.0.137，PC2 的 IP 地址为 192.168.0.136）。

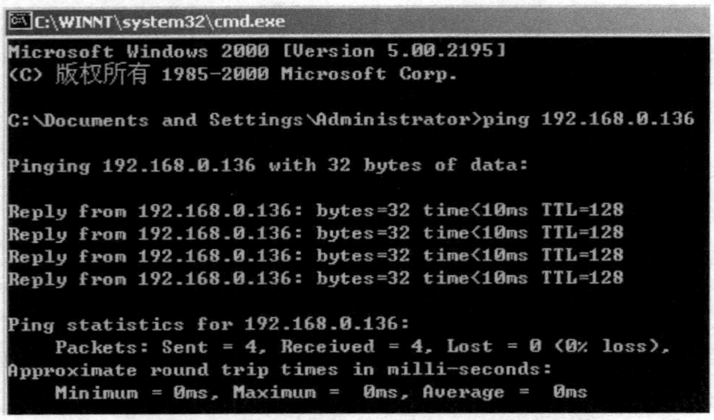

图 7-2　测试骨干网络断链后的连通状况

（2）在 PC1 上继续使用"ping 192.168.0.136 –t"命令，从主机 PC1 上用连续 Ping 命令测试与 PC2 是否连通。

（3）拔掉连接交换机 Switch A 与 Switch B 的 Fa0/1 口之间的连线，观察丢包情况，如

任务 ❼ 使用 RSTP 技术加快网络收敛

图 7-3 所示，显示丢包数为一个。

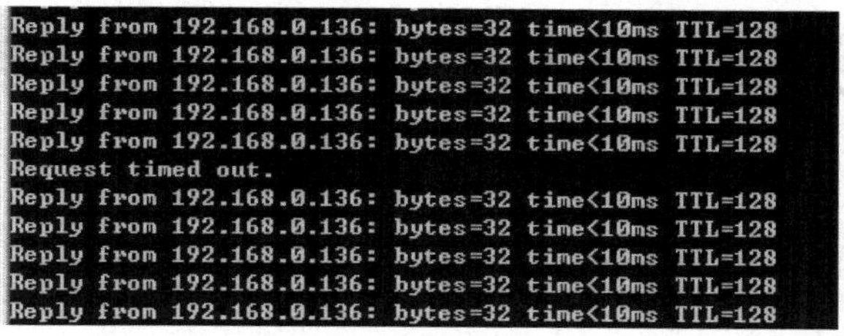

图 7-3 观察丢包情况

【注意事项】

（1）实验时一定要先启用 STP，再按拓扑图连接。

（2）交换机默认关闭 STP，如果网络在物理链路上存在环路，则必须手工开启 STP。

（3）交换机产品默认生成树协议版本为 MSTP，在配置时须注意配置生成树协议的版本。

任务 ❽ 使用 MSTP 技术实现网络负载均衡

【任务描述】

STP 和 RSTP 基于整个交换网络产生一个树形拓扑，所有 VLAN 共享一棵生成树，这种结构不能实现网络中流量的负载均衡，使得网络传输中有些交换设备繁忙，而另一些交换设备又很空闲。为保障网络的稳定性，通常在网络中心使用多台设备实现冗余和备份。实施基于 VLAN 多实例的生成树协议（MSTP）不仅适应了多 VLAN 场景，还在冗余的网络中保障了网络稳健性，并实现了网络的负载均衡。

【任务目标】

配置 MSTP 技术，实现网络负载均衡，了解 MSTP 工作机制。

【组网拓扑】

图 8-1 所示的网络拓扑是某大学的网络中心利用 MSTP 技术，实现网络负载均衡的工作场景。其中，PC1 和 PC3 在 VLAN 10 中，IP 地址分别为 172.16.1.10/24 和 172.16.1.30/24；PC2 在 VLAN 20 中，PC4 在 VLAN 40 中。

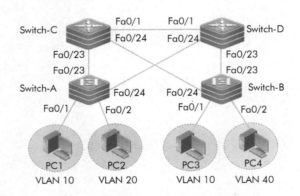

图 8-1　网络中心实施 MSTP 组网拓扑图

（备注：具体的设备连接接口信息可根据实际情况决定）

【设备清单】

二层交换机（两台），三层交换机（两台），网线（若干），PC（若干）。

【关键技术】

MSTP 技术是 STP 和 RSTP 技术的升级，除保留了低版本的特性外，MSTP 还考虑到网络中 VLAN 技术的应用，引入了实例（Instance）和域（Domain）的概念。

任务 ⑧ 使用 MSTP 技术实现网络负载均衡

MSTP 定义了"实例"的概念。这里的实例为网络中多个 VLAN 的组合，把多个 VLAN 捆绑到一个实例中，既可以节省通信开销和资源占用率，又可以针对多个 VLAN 进行生成树运算，而不会阻断网络中应保留的链路，还可以让各个实例的数据经由不同路径转发，实现网络的负载分担。

MSTP 的精妙之处在于把支持 MSTP 的交换机和不支持 MSTP 的交换机划分成不同的区域，分别称为 MST 域和 SST 域。在 MST 域内部运行多实例化的生成树，在 MST 域的边缘运行 RSTP 兼容的内部生成树（Internal Spanning Tree，IST）。

安装在 MST 域内的交换机之间使用 MSTP + BPDU 交换拓扑信息，IST 域内的交换机使用 STP/RSTP/PVST+ BPDU 交换拓扑信息。在 MST 域与 IST 域之间的边缘上，IST 设备会认为对接的设备也是一台 RSTP 设备。而 MST 设备在边缘端口上的状态取决于内部生成树的状态，端口上所有 VLAN 的生成树状态将保持一致。

MSTP 相对于之前的种种生成树协议而言，优势非常明显。MSTP 具有 VLAN 认知能力，设置 VLAN 映射表（即 VLAN 和生成树对应关系表），把 VLAN 和生成树关联起来。MSTP 通过"实例"（将多个 VLAN 整合到一个集合中）这个概念，将多个 VLAN 捆绑到一个实例中，节省通信开销和资源占用率。此外，MSTP 将环路网络修剪成为一个无环的树形网络，避免报文在环路网络中的增生和无限循环，还提供了数据转发的多条冗余路径，在数据转发过程中实现 VLAN 中数据的负载均衡，并可以实现类似 RSTP 的端口状态快速切换效果，最难能可贵的是，MSTP 可以很好地向下兼容 STP/RSTP。

【实施步骤】

备注：配置前使用"show interface status"命令查看接口名称及状态，常用接口有 FastEthernet（百兆）、GigabitEthernet（吉比特）和 TenGigabitEthernet（万兆）。

（1）在二层接入交换机 Switch-A 上划分 VLAN 并配置 Trunk。

```
Ruijie(config)#hostname Switch-A
Switch-A(config)#vlan 10                              // 创建 VLAN
Switch-A(config-vlan)#vlan 20
Switch-A(config-vlan)#vlan 40
Switch-A(config-vlan)#exit

Switch-A(config)#interface FastEthernet 0/1           // 分配接口到指定 VLAN
Switch-A(config-if-FastEthernet 0/1)#switchport access vlan 10
Switch-A(config-if-FastEthernet 0/1)#exit
Switch-A(config)#interface FastEthernet 0/2           // 分配接口到指定 VLAN
Switch-A(config-if-FastEthernet 0/2)#switchport access vlan 20
Switch-A(config-if-FastEthernet 0/2)#exit
Switch-A(config)#interface FastEthernet 0/23          // 设置上连干道端口
Switch-A(config-if-FastEthernet 0/23)#switchport mode trunk
Switch-A(config-if-FastEthernet 0/23)#exit
Switch-A(config)#interface FastEthernet 0/24
```

```
Switch-A(config-if-FastEthernet 0/24)#switchport mode trunk
Switch-A(config-if-FastEthernet 0/24)#exit
Switch-A(config)#
```

（2）在二层接入交换机 Switch-B 上划分 VLAN 并配置 Trunk。

```
Ruijie(config)#hostname Switch-B
Switch-B(config)#vlan 10                              // 创建 VLAN
Switch-B(config-vlan)#vlan 20
Switch-B(config-vlan)#vlan 40
Switch-B(config-vlan)#exit

Switch-B(config)#interface FastEthernet 0/1           // 分配接口到指定 VLAN
Switch-B(config-if-FastEthernet 0/1)#switchport access vlan 10
Switch-B(config-if-FastEthernet 0/1)#exit
Switch-B(config)#interface FastEthernet 0/2           // 分配接口到指定 VLAN
Switch-B(config-if-FastEthernet 0/2)#switchport access vlan 40
Switch-B(config-if-FastEthernet 0/2)#exit
Switch-B(config)#interface FastEthernet 0/23          // 设置上连干道端口
Switch-B(config-if-FastEthernet 0/23)#switchport mode trunk
Switch-B(config-if-FastEthernet 0/23)#exit
Switch-B(config)#interface FastEthernet 0/24
Switch-B(config-if-FastEthernet 0/24)#switchport mode trunk
Switch-B(config-if-FastEthernet 0/24)#exit
```

（3）在三层汇聚交换机 Switch-C 上划分 VLAN 并配置 Trunk。

```
Ruijie(config)#hostname Switch-C
Switch-C(config)#vlan 10                              // 创建 VLAN
Switch-C(config-vlan)#vlan 20
Switch-C(config-vlan)#vlan 40
Switch-B(config-vlan)#exit

Switch-C(config)#interface FastEthernet 0/1           // 设置互连干道端口
Switch-C(config-if-FastEthernet 0/1)#switchport mode trunk
Switch-C(config-if-FastEthernet 0/1)#exit
Switch-C(config)#interface FastEthernet 0/23
Switch-C(config-if-FastEthernet 0/23)#switchport mode trunk
Switch-C(config-if-FastEthernet 0/23)#exit
Switch-C(config)#interface FastEthernet 0/24
Switch-C(config-if-FastEthernet 0/24)#switchport mode trunk
Switch-C(config-if-FastEthernet 0/24)#exit
```

任务 ❽ 使用 MSTP 技术实现网络负载均衡

（4）在三层汇聚交换机 Switch-D 上划分 VLAN 并配置 Trunk。

```
Ruijie(config)#hostname Switch-D
Switch-D(config)#vlan 10              // 创建 VLAN
Switch-D(config-vlan)#vlan 20
Switch-D(config-vlan)#vlan 40
Switch-D(config-vlan)#exit

Switch-D(config)#interface FastEthernet 0/1
Switch-D(config-if-FastEthernet 0/1)#switchport mode trunk
Switch-D(config-if-FastEthernet 0/1)#exit
Switch-D(config-if-FastEthernet 0/23)#interface FastEthernet 0/23
Switch-D(config-if-FastEthernet 0/23)#switchport mode trunk
Switch-D(config-if-FastEthernet 0/23)#exit
Switch-D(config)#interface FastEthernet 0/24
Switch-D(config-if-FastEthernet 0/24)#switchport mode trunk
Switch-D(config-if-FastEthernet 0/24)#exit
```

（5）在二层接入交换机 Switch-A 上配置 MSTP。

```
Switch-A(config)#
Switch-A(config)#spanning-tree                          // 开启生成树协议
Switch-A(config)#spanning-tree mode mstp                // 配置生成树模式为 MSTP
Switch-A(config)#spanning-tree mst configuration        // 进入 MSTP 配置模式
Switch-A(config-mst)#instance 1 vlan 1,10
                                        // 配置 Instance 1 并关联 VLAN 1 和 VLAN 10
Switch-A(config-mst)#instance 2 vlan 20,40
                                        // 配置 Instance 2 并关联 VLAN 20 和 VLAN 40
Switch-A(config-mst)#name region1                       // 配置域名称
Switch-A(config-mst)#revision 1                         // 配置修订号
Switch-A(config-mst)#end
```

验证：在 Switch-A 上验证 MSTP 配置。

```
Switch-A#show spanning-tree mst configuration
Multi spanning tree protocol : Enabled
Name       : region1
Revision : 1
Instance   Vlans Mapped
--------   ------------------------------------------------------------
0          2-9,11-19,21-39,41-4094
1          1,10
2          20,40
```

（6）在二层接入交换机 Switch-B 上配置 MSTP。

35

```
Switch-B(config)#
Switch-B(config)#spanning-tree                              // 开启生成树协议
Switch-B(config)#spanning-tree mode mstp                    // 配置生成树模式为 MSTP
Switch-B(config)#spanning-tree mst configuration            // 进入 MSTP 配置模式
Switch-B(config-mst)#instance 1 vlan 1,10
                                    // 配置 Instance 1 并关联 VLAN 1 和 VLAN 10
Switch-B(config-mst)#instance 2 vlan 20,40
                                    // 配置 Instance 2 并关联 VLAN 20 和 VLAN 40
Switch-B(config-mst)#name region1                           // 配置域名称
Switch-B(config-mst)#revision 1                             // 配置修订号
Switch-B(config-mst)#end
```

验证：在 Switch-B 上验证 MSTP 配置。

```
Switch-B#show spanning-tree mst configuration
Multi spanning tree protocol : Enabled
Name       : region1
Revision : 1
Instance   Vlans Mapped
--------   ------------------------------------------------------------
0          2-9,11-19,21-39,41-4094
1          1,10
2          20,40
```

（7）在三层汇聚交换机 Switch-C 上配置 MSTP。

```
Switch-C(config)#spanning-tree                              // 开启生成树协议
Switch-C(config)#spanning-tree mode mstp                    // 配置生成树模式为 MSTP
Switch-C(config)#spanning-tree mst 1 priority 4096
        // 配置交换机 Switch-C 在 Instance 1 中的优先级为 4096，成为 Instance 1 的根
Switch-C(config)#spanning-tree mst configuration            // 进入 MSTP 配置模式
Switch-C(config-mst)#instance 1 vlan 1,10
                                    // 配置 Instance 1 并关联 VLAN 1 和 VLAN 10
Switch-C(config-mst)#instance 2 vlan 20,40
                                    // 配置 Instance 2 并关联 VLAN 20 和 VLAN 40
Switch-C(config-mst)#name region1                           // 配置域名为 region1
Switch-C(config-mst)#revision 1                             // 配置修订号
Switch-C(config-mst)#end
```

验证：在 Switch-C 上验证 MSTP 配置。

```
Switch-C#show spanning-tree mst configuration
Multi spanning tree protocol : Enabled
Name       : region1
Revision : 1
```

任务 ❽　使用 MSTP 技术实现网络负载均衡

```
Instance    Vlans Mapped
--------    ------------------------------------------------
0           2-9,11-19,21-39,41-4094
1           1,10
2           20,40
```

（8）在三层汇聚交换机 Switch-D 上配置 MSTP。

```
Switch-D(config)#
Switch-D(config)#spanning-tree                        // 开启生成树协议
Switch-D(config)#spanning-tree mode mstp              // 配置生成树模式为 MSTP
Switch-D(config)#spanning-tree mst 2 priority 4096
        // 配置交换机 Switch-D 在 Instance 2 优先级为 4096，因为 Instance 2 成为根
Switch-D(config)#spanning-tree mst configuration      // 进入 MSTP 配置模式
Switch-D(config-mst)#instance 1 vlan 1,10
                     // 配置 Instance 1 并关联 VLAN 1 和 VLAN 10
Switch-D(config-mst)#instance 2 vlan 20,40
                     // 配置 Instance 2 并关联 VLAN 20 和 VLAN 40
Switch-D(config-mst)#name region1                     // 配置域名为 region1
Switch-D(config-mst)#revision 1                       // 配置修订号
Switch-D(config-mst)#end
```

【测试验证】

（1）在交换机 Switch-D 上验证 MSTP 配置。

```
Switch-D#show spanning-tree mst configuration
Multi spanning tree protocol : Enabled
Name       : region1
Revision : 1
Instance    Vlans Mapped
--------    ------------------------------------------------
0           2-9,11-19,21-39,41-4094
1           1,10
2           20,40
```

（2）在三层汇聚交换机 Switch-C 上查看 MSTP 选举结果。

```
Switch-C#show spanning-tree mst 1
MST 1 vlans mapped : 1,10
BridgeAddr : 00d0.f8ff.4e3f
Priority : 4096
TimeSinceTopologyChange : 0d:7h:21m:17s
TopologyChanges : 0
DesignatedRoot : 100100D0F8FF4E3F       // Switch-C 是 Instance 1 的生成树的根
```

多层交换技术（实践篇）

```
RootCost : 0
RootPort : 0
```
// 从上述 show 命令的输出结果可以看出，交换机 Switch-C 为 Instance 1 中的根交换机

（3）在三层汇聚交换机 Switch-D 上查看 MSTP 选举结果。

```
Switch-D#show spanning-tree mst 2
MST 2 vlans mapped : 20,40
BridgeAddr : 00d0.f8ff.4662
Priority : 4096
TimeSinceTopologyChange : 0d:7h:31m:0s
TopologyChanges : 0
DesignatedRoot : 100200D0F8FF4662      // Switch-D 是 Instance 2 的生成树的根
RootCost : 0
RootPort : 0
```
// 从上述 show 命令的输出结果可以看出，交换机 Switch-D 为 Instance 2 中的根交换机

（4）在二层接入交换机 Switch-A 上查看 MSTP 选举结果。

```
Switch-A#show spanning-tree mst 1
MST 1 vlans mapped : 1,10
BridgeAddr : 00d0.f8fe.1e49
Priority : 32768
TimeSinceTopologyChange : 7d:3h:19m:31s
TopologyChanges : 0
DesignatedRoot : 100100D0F8FF4E3F  // Instance 1 生成树的根交换机是 Switch-C
RootCost : 200000
RootPort : Fa0/23
```
/* 从上述 show 命令的输出结果可以看出，在 Instance 1 中，交换机 Switch-A 的 Fa0/23 端口为根端口，因此 VLAN 1 和 VLAN 10 的 IP 数据包经端口 Fa0/23 转发 */

```
Switch-A#show spanning-tree mst 2
MST 2 vlans mapped : 20,40
BridgeAddr : 00d0.f8fe.1e49
Priority : 32768
TimeSinceTopologyChange : 7d:3h:19m:31s
TopologyChanges : 0
DesignatedRoot : 100200D0F8FF4662 // Instance 2 生成树的根交换机是 Switch-D
RootCost : 200000
RootPort : Fa0/24
```
/* 从上述 show 命令的输出结果可以看出，在 Instance 2 中，交换机 Switch-A 的 Fa0/24 端口为根端口，因此 VLAN20 和 VLAN40 的 IP 数据包经端口 Fa0/24 转发 */

任务 ⑧　使用 MSTP 技术实现网络负载均衡

【注意事项】

（1）划分在同一个域中的各台交换机，须配置相同的域名、相同的修订号、相同的 Instance VLAN 映射表。

（2）交换机中，Instance 0 是默认实例，强制存在，其他实例可以创建和删除。

（3）若要将整个生成树恢复为默认状态，则必须使用"spanning-tree reset"命令。

任务 ⑨ VRRP 实现单备份组网络出口冗余 1

【任务描述】

某公司在全国各地都建有分公司，建设有互连互通的企业网。其中，在天津分公司的网络建设初期，网络出口只有一条出口光纤线路。在使用的过程中，单链路经常出现线路故障，导致网络中断。为了保障天津分公司网络出口的稳定性，希望在天津分公司的出口网络中增加一条出口线路，通过 VRRP 技术使两条出口线路互为备份。

【任务目标】

掌握单备份组的 VRRP 技术，使用 VRRP 实现出口冗余。

【组网拓扑】

图 9-1 所示为天津分公司连接到北京总公司的网络场景。天津分公司的出口网络中使用两台路由器作为网络出口，在路由器上配置 VRRP 技术实现出口网络冗余，通过两条线路连接到 Internet 中，两条出口线路互为备份。

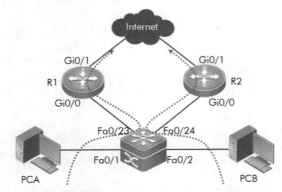

图 9-1 天津分公司连接到北京总公司的网络场景
（备注：具体的设备连接接口信息可根据实际情况决定）

【设备清单】

路由器（两台），交换机（1 台），网线（若干），PC（若干）。

【关键技术】

通常，一个网络内所有主机会设置一个默认网关，这样网络内的主机发到外网的 IP 报文将通过默认网关发到出口路由器，实现主机与外网的通信。

当本地网络中的出口路由器掉线时，本网段内所有以该路由器为默认路由（网关）

任务 ⑨　VRRP 实现单备份组网络出口冗余 1

的主机将会失去和外网的通信，从而产生单点故障。VRRP 技术就是为解决上述问题而提出的。

VRRP 是一种路由冗余协议，它可以把一台虚拟路由器的 IP 地址动态分配到局域网中 VRRP 路由器组中的任意一台上。其中，控制虚拟路由器 IP 地址的 VRRP 路由器称为主路由器，它负责转发数据包。一旦主路由器不可用，则 VRRP 提供动态的转移机制，允许备份路由器承担转发功能。

VRRP 是一种路由容错协议，路由器上开启 VRRP 功能后，会根据优先级确定自己在备份组中的角色。优先级高的路由器成为主路由器，优先级低的成为备份路由器。主路由器定期发送 VRRP 通告报文，通知备份组内的其他路由器自己工作正常；备份路由器则启动定时器等待通告报文的到来。

VRRP 在不同的主备抢占方式下，主备角色的替换方式不同。

在抢占方式下，当主路由器收到 VRRP 通告报文后，会将自己的优先级与通告报文中的优先级进行比较，如果大于通告报文中的优先级，则成为主路由器，否则将保持备份状态。

在非抢占方式下，只要主路由器没有出现故障，则备份组中的路由器始终保持备份状态。备份组中的路由器随后即使配置了更高的优先级，也不会成为主路由器。

如果备份路由器的定时器超时后，仍未收到主路由器发送的 VRRP 通告报文，则认为主路由器已经掉线。此时，备份路由器会认为自己是主路由器，开始对外发送 VRRP 通告报文。

【实施步骤】

（1）在出口路由器 R1 上配置 VRRP。

```
Ruijie#configure terminal
Ruijie(config)#interface GigabitEthernet 0/0
Ruijie(config-if-GigabitEthernet 0/0)#ip address 192.168.1.1 255.255.255.0
                                                        // 配置物理接口 IP 地址
Ruijie(config-if-GigabitEthernett 0/0)#vrrp 1 ip 192.168.1.254
                                                        // 指定 VRRP 虚拟地址
Ruijie(config-if-GigabitEthernet 0/0)#vrrp 1 priority 120
                        // 指定该接口 VRRP 的优先级，越大越优先，默认 100
Ruijie(config-if-GigabitEthernet 0/0)#vrrp 1 track GigabitEthernet 0/1 30
                        // 检测上连接口 Fa0/1 掉线后优先级降低 30，切换为备份网关
Ruijie(config-if-GigabitEthernet 0/0)#exit
```

（2）在出口路由器 R2 上配置 VRRP。

```
Ruijie#configure terminal
Ruijie(config)#interface GigabitEthernet 0/0
Ruijie(config-if-GigabitEthernet 0/0)#ip address 192.168.1.2 255.255.255.0
Ruijie(config-if-GigabitEthernet 0/0)#vrrp 1 ip 192.168.1.254
```

多层交换技术（实践篇）

```
                                        // 指定 VRRP 虚拟地址
Ruijie(config-if-GigabitEthernet 0/0)#end
```

【测试验证】

（1）使用"show vrrp brief"命令查看 VRRP 协商状态。

```
Ruijie#show vrrp brief
Interface Grp      Pri   timer Own  Pre State   Master addr  Group addr
GigabitEthernet0/0  1     90   3  -  P  Backup  192.168.1.2  192.168.1.254
```

（2）使用"show vrrp 1"命令查看 VRRP 1 的信息。

```
Ruijie#show vrrp 1
GigabitEthernet    0/0 -    Group    1
State is Backup
Virtual IP address is 192.168.1.254  configured
Virtual MAC address is  0000.5e00.0101
Advertisement interval is 1 sec
Preemption is enabled
min delay is 0 sec
Priority is    90
Master Router is 192.168.1.2 , priority is 100
Master Advertisement interval is 1 sec
Master Down interval is 3 sec
Tracking    state of 1 interface, 0 up:
down GigabitEthernet 0/1 priority decrement=30
```

（3）分别给分公司内网中的测试计算机 PCA 和 PCB 配置"192.168.1.0/24"网段的 IP 地址，网关配置为虚拟 VRRP 网关 IP 地址 192.168.1.254，如图 9-2 所示。

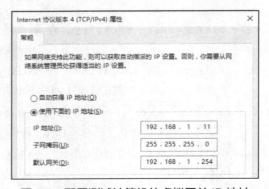

图 9-2　配置测试计算机的虚拟网关 IP 地址

在测试计算机 PCA 上，使用 Ping 命令测试虚拟网关，以及 PCA 与 PCB 的连通状况，测试结果如图 9-3 所示。

任务 ❾　VRRP 实现单备份组网络出口冗余 1

图 9-3　测试结果

任务 ⑩ VRRP 实现单备份组网络出口冗余 2

【任务描述】

某集团公司总公司在北京,分公司在上海,分公司和总公司之间通过一条专用线路实现连接,在网络传输过程中,经常出现由于线路故障导致网络中断的情况。

为了保障分公司与总公司之间网络的稳定性,希望在分公司的出口网络中增加一条备份出口链路,配置 VRRP 技术通过两条冗余线路连接到总公司,两条线路互为备份。

【任务目标】

配置 VRRP 单备份组,实现 VRRP 的冗余备份。

【组网拓扑】

图 10-1 所示网络拓扑为某分公司采用单链路专线接入总公司网络场景,但容易出现单点故障。可以增加一条备份线路接入总公司网络,并在分公司的出口路由器上运行 VRRP 技术,两条线路互为备份,出现故障时自动切换。

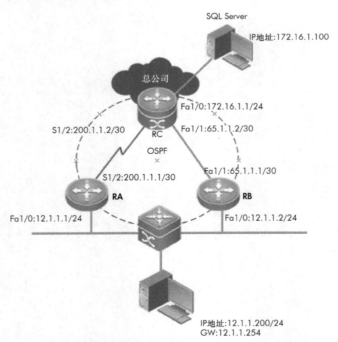

图 10-1 某分公司采用单链路专线接入总公司网络场景

(备注:具体的设备连接端口信息可根据实际情况决定,本任务采用 WAN 接口)

任务 ⑩　VRRP 实现单备份组网络出口冗余 2

【设备清单】

路由器（3 台），交换机（1 台），网线（若干），PC（若干）。

【关键技术】

VRRP 技术是解决网络中主机配置单网关容易出现单点故障问题的技术，即将多台出口路由器加入到一个 VRRP 组中，每一个 VRRP 组虚拟出一台虚拟路由器，作为内部网络中主机的网关。在一个 VRRP 组中，所有物理路由器中优先级最高的被选举为主路由器，传输到虚拟路由器上的数据都由主路由器转发。当主路由器出现故障时，备份路由器成为主路由器，承担虚拟路由器数据转发工作，从而保证出口网络的稳健性。

【实施步骤】

（1）在出口路由器 RA 上配置网络出口基本信息。

```
Ruijie#config terminal
Ruijie(config)#hostname RA
RA(config)#interface serial 1/0
RA(config-if-Serial 1/0)#ip address 200.1.1.1 255.255.255.252
RA(config-if-Serial 1/0)#exit
RA(config)#interface FastEthernet 0/0
RA(config-if-FastEthernet 0/0)#ip address 12.1.1.1 255.255.255.0
RA(config-if-FastEthernet 0/0)# exit
RA(config)#
```

（2）在出口路由器 RB 上配置网络出口基本信息。

```
Ruijie#config terminal
Ruijie(config)#hostname RB
RB(config)#interface FastEthernet 0/0
RB(config-if-FastEthernet 0/0)#ip address 12.1.1.2 255.255.255.0
RB(config-if-FastEthernet 0/0)#exit
RB(config)#interface FastEthernet 0/1
RB(config-if-FastEthernet 0/1)#ip address 65.1.1.1 255.255.255.252
RB(config-if-FastEthernet 0/1)#exit
RB(config)#
```

（3）在总公司接入路由器 RC 上配置接入网络基本信息。

```
Ruijie#config terminal
Ruijie(config)#hostname RC
RC(config)#interface serial 1/0
RC(config-if-Serial 1/0)#ip address 200.1.1.2 255.255.255.252
RC(config-if-Serial 1/0)#clock rate 64000
RC(config-if-Serial 1/0)#exit
RC(config)#interface FastEthernet 0/1
RC(config-if-FastEthernet 0/1)#ip address 65.1.1.2 255.255.255.252
```

```
RC(config-if-FastEthernet 0/1)#exit
RC(config)#interface FastEthernet 0/0
RC(config-if-FastEthernet 0/0)#ip address 172.16.1.1 255.255.255.0
RC(config-if-FastEthernet 0/0)#exit
```

（4）在全网路由器上配置 OSPF 路由，实现全网连通。

```
RA(config)#router ospf
RA(config-router)#network 12.1.1.0 0.0.0.255 area 0
RA(config-router)#network 200.1.1.0 0.0.0.3 area 0
RA(config-router)#exit


RB(config)#router ospf
RB(config-router)#network 12.1.1.0 0.0.0.255 area 0
RB(config-router)#network 65.1.1.0 0.0.0.3 area 0
RB(config-router)#exit


RC(config)#router ospf
RC(config-router)#network 65.1.1.0 0.0.0.3 area 0
RC(config-router)#network 172.16.1.0 0.0.0.255 area 0
RC(config-router)#network 200.1.1.0 0.0.0.3 area 0
RC(config-router)#exit
```

（5）在分公司出口路由器 RA 上配置 VRRP，实现网络出口冗余和备份。

```
RA(config)#interface FastEthernet 0/0
RA(config-if-FastEthernet 0/0)#vrrp 32 ip 12.1.1.254
                            // 指定 VRRP 虚拟地址为 12.1.1.254
RA(config-if-FastEthernet 0/0)#vrrp 32 priority 120
      // 将 RA 在 VRRP 组 32 中的优先级配置为较高的 120，成为主路由器，值越大，优先级越高
RA(config-if-FastEthernet 0/0)#vrrp 32 track serial 1/0 30
      /* 设置 RA 在 VRRP 组 32 中对端口 S1/2 进行监控，当监控的端口状态为 DOWN 时，路由
器优先级降低 30，切换为备份路由器 */
RA(config-if-FastEthernet 0/0)#exit
```

（6）在分公司出口路由器 RB 上配置 VRRP，实现网络出口冗余和备份。

```
RB(config)#interface FastEthernet 0/0
RB(config-if-FastEthernet 0/0)#vrrp 32 ip 12.1.1.254
                            // 指定 VRRP 虚拟地址为 12.1.1.254，默认优先级为 100
RB(config-if-FastEthernet 0/0)#exit
```

（7）在分公司网络出口路由器上验证 VRRP 状态。

① 使用"show vrrp brief"命令验证配置。

```
RA#show vrrp brief
Interface          Grp  Pri Time  Own Pre State   Master addr    Group addr
FastEthernet1/0    32   120  -     -   P  Master  12.1.1.1       12.1.1.254
```

任务 ⑩ VRRP 实现单备份组网络出口冗余 2

/* 从 show 命令的输出结果可以看到，RA 路由器在 VRRP 组 32 中，优先级为 120，状态为主路由器 */

② 关闭总公司路由器 RC 的端口 S1/2，观察网络连接信息。

```
RC(config)#
RC(config)#interface serial 1/2
RC(config-if-serial 1/2)#shutdown        // 用 shutdown 命令关闭端口
RC(config-if-serial 1/2)#exit
RC(config)#
```

③ 查看分公司的出口路由器 RA 的 S1/2 端口状态，其也变为 DOWN。

```
RA(config)#
%LINK CHANGED: Interface serial 1/2, changed state to down
%LINE PROTOCOL CHANGE: Interface serial 1/2, changed state to DOWN
```

```
RA#show ip interface brief       // 在分公司路由器上查看端口状态
Interface            IP-Address(Pri)      OK       Status
serial 1/2           200.1.1.1/30         YES      DOWN
serial 1/3           no address           YES      DOWN
FastEthernet 1/0     12.1.1.1/24          YES      UP
FastEthernet 1/1     no address           YES      DOWN
Null 0               no address           YES      UP
```

④ 在分公司出口路由器 RA 上观察 VRRP 信息。

```
RA#show vrrp brief    // 使用 "show vrrp brief" 命令验证配置
Interface         Grp  Pri  Time  Own  Pre  State   Master addr  Group addr
FastEthernet 1/0  32   90   -     -    P    Backup  12.1.1.2     12.1.1.254
```

/* 从 show 命令的输出结果可以看到，当监控端口状态变为 DOWN 时，路由器 RA 在 VRRP 组 32 中，优先级为 90，状态为备份路由器 */

（8）在分公司网络中测试网络配置。

首先，分别给分公司内网中的测试计算机配置内网 IP 地址，配置虚拟 VRRP 网关地址为 12.1.1.254/24。其次，在分公司内网中的测试计算机上，使用 Ping 命令测试虚拟网关及其他计算机之间的连通情况。

任务 11 VRRP 实现多备份组冗余网络出口

【任务描述】

某公司的天津分公司和北京总公司之间只有一条出口专线,经常出现由于单链路故障而导致网络中断的情况。

为了保障分公司与总公司之间网络连接的稳定性,希望在分公司的出口网络中增加一条出口专线,使用两条线路连接到总公司网络,并通过配置 VRRP 技术使两条线路互为备份,实现出口网络中数据流量的负载均衡。

【任务目标】

配置 VRRP 的多备份组功能,增强出口网络稳健性,实现 VRRP 的负载均衡模式。

【组网拓扑】

图 11-1 所示为天津分公司连接北京总公司网络场景,天津分公司通过两条专线连接到总公司网络,通过 VRRP 技术使两条线路互为备份,在保证网络稳定运行的同时实现网络中流量的负载均衡。

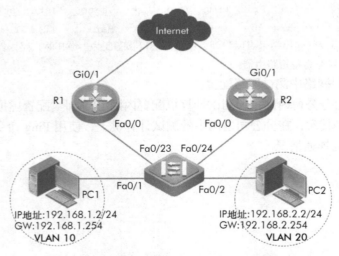

图 11-1 天津分公司连接北京总公司网络场景

备注 1:具体设备连接接口信息可根据实际情况决定。

备注 2:使用两台路由器与一台交换机搭建网络环境,在路由器上配置 VRRP。其中,R1 为 192.168.1.0 网络的主网关,R2 为 192.168.2.0 网络的主网关。

任务 ⑪ VRRP 实现多备份组冗余网络出口

【设备清单】

路由器（两台），交换机（1台），网线（若干），PC（若干）。

【实施步骤】

（1）在分公司出口路由器 R1 上配置接口信息和 VRRP。

① 创建 VLAN 10 网络中的用户出口。

```
Ruijie#configure terminal
Ruijie(config)#hostname R1
R1(config)#interface FastEthernet 0/0.1                  // 创建子接口 1
R1(config-if-FastEthernet 0/0.1)#encapsulation dot1Q 10
                                    // 子接口 1 封装在 VLAN 10 中
R1(config-if-FastEthernet 0/0.1)#ip address 192.168.1.1 255.255.255.0
R1(config-if-FastEthernet 0/0.1)#vrrp 1 ip 192.168.1.254
                                    // 指定 VRRP1 虚拟地址
R1(config-if-FastEthernet 0/0.1)#vrrp 1 priority 120
            // 指定该接口 VRRP1 的优先级，值越大，优先级越高，默认为 100
R1(config-if-FastEthernet 0/0.1)#vrrp 1 track FastEthernet 0/1 30
            // 检测上行接口 Fa0/0 在掉线后优先级降 30，切换为备份网关
R1(config-if-FastEthernet 0/0.1)#exit
```

② 创建 VLAN 20 网络中的用户出口。

```
R1(config)#interface FastEthernet 0/0.2                  // 创建子接口 2
R1(config-if-FastEthernet 0/0.2)#encapsulation dot1Q 20
                                    // 子接口 2 封装在 VLAN 20 中
R1(config-if-FastEthernet 0/0.2)#ip address 192.168.2.1 255.255.255.0
R1(config-if-FastEthernet 0/0.2)#vrrp 2 ip 192.168.2.254
                                    // 指定 VRRP2 虚拟地址
R1(config-if-FastEthernet 0/0.2)#exit
```

（2）在分公司出口路由器 R2 上配置接口信息和 VRRP。

① 创建 VLAN 10 网络中的用户出口。

```
Ruijie#configure terminal
Ruijie(config)#hostname R2
R2(config)#interface FastEthernet 0/0.1                  // 创建子接口 1
R2(config-if-FastEthernet 0/0.1)#encapsulation dot1Q 10
                                    // 将子接口 1 封装在 VLAN 10 中
R2(config-if-FastEthernet 0/0.1)#ip address 192.168.1.2 255.255.255.0
R2(config-if-FastEthernet 0/0.1)#vrrp 1 ip 192.168.1.254
                                    // 指定 VRRP1 虚拟地址
R2(config-if-FastEthernet 0/0.1)#exit
```

② 创建 VLAN 20 网络中的用户出口。

```
R2(config)#interface FastEthernet 0/0.2                  // 创建子接口 2
R2(config-if-FastEthernet 0/0.2)#encapsulation dot1Q 20
                                    // 将子接口 2 封装在 VLAN 20 中
```

多层交换技术（实践篇）

```
R2(config-if-FastEthernet 0/0.2)#ip address 192.168.2.2 255.255.255.0
R2(config-if-FastEthernet 0/0.2)#vrrp 2 ip 192.168.2.254  // 指定VRRP2 虚
拟地址
R2(config-if-FastEthernet 0/0.2)#vrrp 2 priority 120
R2(config-if-FastEthernet 0/0.2)#vrrp 2 track FastEthernet 0/1 30
                  // 检测上行接口 Fa0/0 在掉线后优先级降 30，切换为备份网关
R2(config-if-FastEthernet 0/0.2)#exit
```

（3）在分公司接入交换机上配置用户接入信息。

```
Ruijie#configure terminal
Ruijie(config)#interface range FastEthernet 0/23-24
Ruijie(config-if-range)#switchport mode trunk
Ruijie(config-if-range)#exit

Ruijie(config)#vlan 10
Ruijie(config-vlan)#vlan 20
Ruijie(config-vlan)#exit

Ruijie(config)#interface FastEthernet 0/1
Ruijie(config-if-FastEthernet 0/1)#switch access vlan 10
Ruijie(config-if-FastEthernet 0/1)#exit
Ruijie(config)#interface FastEthernet 0/2
Ruijie(config-if-fastEthernet 0/2)#switch access vlan 20
Ruijie(config-if-fastEthernet 0/2)#exit
Ruijie(config)#
```

（4）在分公司网络出口路由器上验证 VRRP 配置信息。

① 在分公司出口路由器 R1 上，使用 "show vrrp brief" 命令查看 VRRP 协商信息。

如图 11-2 所示，出口路由器 R1 的主网关 IP 地址为 192.168.1.254/24；备份网关 IP 地址为 192.168.2.254/24。

```
Ruijie#show vrrp br
Ruijie#show vrrp brief
Interface            Grp  Pri  timer  Own  Pre  State   Master addr
   Group addr
Gi0/0.1               1   120  3.53    -    P   Master  192.168.1.1
   192.168.1.254
Gi0/0.2               2   100  3.60    -    P   Master  192.168.2.1
   192.168.2.254
```

图 11-2　在 R1 上查看 VRRP 协商信息

② 在出口路由器 R1 上，使用 "show vrrp 1" 命令查看 VRRP1 状态信息，如图 11-3 所示。

```
Ruijie#show vrrp 1
GigabitEthernet 0/0.1 - Group 1
  State is Master
  Virtual IP address is 192.168.1.254 configured
  Virtual MAC address is 0000.5e00.0101
  Advertisement interval is 1 sec
  Preemption is enabled
    min delay is 0 sec
  Priority is 120
  Master Router is 192.168.1.1 (local), priority is 120
  Master Advertisement interval is 1 sec
  Master Down interval is 3.53 sec
```

图 11-3　在 R1 上查看 VRRP1 状态信息

任务 ⑪　VRRP 实现多备份组冗余网络出口

③ 在出口路由器 R2 上，使用 "show vrrp brief" 命令查看 VRRP 协商信息，如图 11-4 所示。路由器 R2 主网关 IP 地址为 192.168.2.254/24；备份网关 IP 地址为 192.168.1.254/24。

```
Ruijie(config)#show vrrp brief
Interface        Grp   Pri   timer   Own   Pre   State    Master addr      Group addr
Gi0/0.1          1     100   3.60    -     P     Backup   192.168.1.1      192.168.1.254
Gi0/0.2          2     120   3.53    -     P     Master   192.168.2.2      192.168.2.254
```

图 11-4　在 R2 上查看 VRRP 协商信息

④ 在出口路由器 R2 上，使用 "show vrrp 2" 命令查看 VRRP 2 状态信息，如图 11-5 所示。

```
Ruijie#show vrrp 2
GigabitEthernet 0/0.2 - Group 2
  State is Master
  Virtual IP address is 192.168.2.254 configured
  Virtual MAC address is 0000.5e00.0102
  Advertisement interval is 1 sec
  Preemption is enabled
    min delay is 0 sec
  Priority is 120
  Master Router is 192.168.2.2 (local), priority is 120
  Master Advertisement interval is 1 sec
  Master Down interval is 3.53 sec
```

图 11-5　在 R2 上查看 VRRP 2 状态信息

（5）在分公司网络中测试网络连通性。

分别给分公司内网中的测试计算机配置相应内网段的 IP 地址，配置相应虚拟 VRRP 网关。

在分公司内网中的测试计算机上，使用 Ping 命令测试虚拟网关及其他计算机之间的连通情况。

任务 ⑫ 使用 VRRP 技术实现核心网络冗余备份(基于 SVI)

【任务描述】

某集团公司在北京的总部网络承担了连接全国各地分公司网络的任务。总部网络中心采用多台万兆交换机,内部网络按照业务规划有 4 个部门 VLAN。为了增强总部核心网络的稳定性,要求各个部门 VLAN 之间不仅需要在三层上实现各个 SVI 之间的互连互通,还需要配置 VRRP 备份组,实现网关冗余。

【任务目标】

配置基于 SVI 的 VRRP 多备份组,实现基于 SVI 的 VRRP 负载均衡模式。

【组网拓扑】

图 12-1 所示网络拓扑为某集团公司核心网络部署场景,为保障核心网络的稳健性,通过 VRRP 技术对不同部门 VLAN 建立多个 VRRP 备份组,基于 VLAN 实现网关冗余及网络的负载均衡。

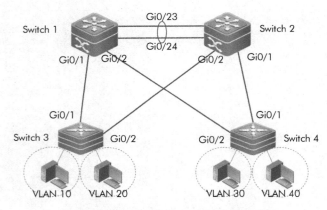

图 12-1 某集团公司核心网络部署场景
(备注:具体的设备连接接口信息可根据实际情况决定)

【设备清单】

二层交换机(两台),三层交换机(两台),网线(若干),PC(若干)。

【关键技术】

VRRP 技术用来为网关设备提供冗余备份,将承担网关功能的一组设备加入到备份组中,形成一台虚拟路由器,局域网内的主机将此虚拟路由器上的 IP 地址设置为默认网关。

任务 ⑫ 使用 VRRP 技术实现核心网络冗余备份（基于 SVI）

VRRP 组中三层路由设备根据优先级，从备份组中选举一台优先级高的设备作为主路由器，负责转发局域网内主机与外部网络的通信流量，其他三层路由设备作为备份路由器。

当主路由器出现故障后，VRRP 重新选举新的主路由器，保证通信流量转发不会中断。

此外，VRRP 可以监视上行端口或链路的状态。当三层设备的上行端口或链路出现故障时，该三层设备主动降低自己的优先级，使得备份组内其他三层设备的优先级高于这台三层设备，承担流量转发任务。

本案例中基于 VLAN 建立不同的 VRRP 组，将三层交换机划分到各个 VRRP 组中；同一台三层交换机在不同的 VRRP 组中承担不同的角色，使得所有三层交换机都承担数据转发任务。

【实施步骤】

（1）分别在 4 台交换机上创建 VLAN。

① 在三层核心交换机 Switch 1 上配置部门 VLAN 信息。

```
Ruijie(config)#hostname Switch 1         // 在交换机 Switch 1 上创建部门接入 VLAN
Switch 1(config)#VLAN 10
Switch 1(config-vlan)#exit
Switch 1(config)#VLAN 20
Switch 1(config-vlan)#exit
Switch 1(config)#VLAN 30
Switch 1(config-vlan)#exit
Switch 1(config)#VLAN 40
Switch 1(config-vlan)#exit
```

② 在三层核心交换机 Switch 2 上配置部门 VLAN 信息。

```
Ruijie(config)#hostname Switch 2         // 在交换机 Switch 2 上创建部门接入 VLAN
Switch 2(config)#VLAN 10
Switch 2(config-vlan)#exit
Switch 2(config)#VLAN 20
Switch 2(config-vlan)#exit
Switch 2(config)#VLAN 30
Switch 2(config-vlan)#exit
Switch 2(config)#VLAN 40
Switch 2(config-vlan)#exit
```

③ 在二层接入交换机 Switch 3 上配置部门 VLAN 信息。

```
Ruijie(config)#hostname Switch 3         // 在交换机 Switch 3 上创建部门 VLAN
Switch 3(config)#VLAN 10
Switch 3(config-vlan)#exit
Switch 3(config)#VLAN 20
Switch 3(config-vlan)#exit
```

④ 在二层接入交换机 Switch 4 上配置部门 VLAN 信息。

```
Ruijie(config)#hostname Switch 4         // 在交换机 Switch 4 上创建部门 VLAN
Switch 4(config)#VLAN 30
```

多层交换技术（实践篇）

```
Switch 4(config-vlan)#exit
Switch 4(config)#VLAN 40
Switch 4(config-vlan)#exit
```

（2）分别在两台核心交换机上打开 SVI，配置 IP 地址。

① 在三层核心交换机 Switch 1 上配置部门 VLAN 的 SVI 地址。

```
Switch 1(config)#interface VLAN 10
Switch 1(config-if-vlan 10)#ip address 192.168.10.1 255.255.255.0
Switch 1(config-if-vlan 10)#exit
Switch 1(config)#interface VLAN 20
Switch 1(config-if-vlan 20)#ip address 192.168.20.1 255.255.255.0
Switch 1(config-if-vlan 20)#exit
Switch 1(config)#interface VLAN 30
Switch 1(config-if-vlan 30)#ip address 192.168.30.1 255.255.255.0
Switch 1(config-if-vlan 30)#exit
Switch 1(config)#interface VLAN 40
Switch 1(config-if-vlan 40)#ip address 192.168.40.1 255.255.255.0
Switch 1(config-if-vlan 40)#exit
```

② 在三层核心交换机 Switch 2 上配置部门 VLAN 的 SVI 地址。

```
Switch 2(config)#interface VLAN 10
Switch 2(config-if-vlan 10)#ip address 192.168.10.2 255.255.255.0
Switch 2(config-if-vlan 10)#exit
Switch 2(config)#interface VLAN 20
Switch 2(config-if-vlan 20)#ip address 192.168.20.2 255.255.255.0
Switch 2(config-if-vlan 20)#exit
Switch 2(config)#interface VLAN 30
Switch 2(config-if-vlan 30)#ip address 192.168.30.2 255.255.255.0
Switch 2(config-if-vlan 30)#exit
Switch 2(config)#interface VLAN 40
Switch 2(config-if-vlan 40)#ip address 192.168.40.2 255.255.255.0
Switch 2(config-if-vlan 40)#exit
```

（3）在核心交换机上配置互连设备的干道链路及聚合端口。

① 在三层核心交换机 Switch 1 上配置干道链路及聚合端口。注意：分配到各个 VLAN 中的端口，根据实验现场的具体连接情况决定。

```
Switch 1(config)#interface range GigabitEthernet 0/1-2    // 配置上连干道链路
Switch 1(config-if-range)#switchport mode trunk
Switch 1(config-if-range)#exit
Switch 1(config)#interface range GigabitEthernet 0/23-24  // 配置核心聚合链路
Switch 1(config-if-range)#port-group 1                    // 核心链路聚合端口
Switch 1(config-if-range)#exit
Switch 1(config)#interface aggregatePort 1                // 把聚合端口配置成干道模式
```

任务 ⑫ 使用 VRRP 技术实现核心网络冗余备份（基于 SVI）

```
Switch 1(config-if-aggregatePort 1)#switchport mode trunk
Switch 1(config-if-aggregatePort 1)#exit
```

② 在三层核心交换机 Switch 2 上配置干道链路及聚合端口

```
Switch 2(config)#interface range GigabitEthernet 0/1-2      // 配置上连干道链路
Switch 2(config-if-range)#switchport mode trunk
Switch 2(config-if-range)#exit
Switch 2(config)#interface range GigabitEthernet 0/23-24    // 配置核心聚合链路
Switch 2(config-if-range)#port-group 1                      // 核心链路聚合端口
Switch 2(config-if-range)#exit
Switch 2(config)#interface aggregatePort 1                  // 把聚合端口配置成干道模式
Switch 2(config-if-aggregatePort 1)#switchport mode trunk
Switch 2(config-if-aggregatePort 1)#exit
```

（4）在接入交换机上配置干道链路及 VLAN 分配信息。

① 在二层接入交换机 Switch 3 上配置干道链路及 VLAN 分配信息。注意：分配到各个 VLAN 中的端口，根据实验现场的具体连接情况决定。

```
Switch 3(config)#interface range GigabitEthernet 0/1-2      // 配置干道链路
Switch 3(config-if-range)#switchport mode trunk
Switch 3(config-if-range)#exit
Switch 3(config)#interface range GigabitEthernet 0/3-14     // 分配部门VLAN 10
Switch 3(config-if-range)#switchport access vlan 10
Switch 3(config-if-range)#exit
Switch 3(config)#interface range GigabitEthernet 0/15-24    // 分配部门VLAN 20
Switch 3(config-if-range)#switchport access vlan 20
Switch 3(config-if-range)#exit
Switch 3(config)#
```

② 在二层接入交换机 Switch 4 上配置干道链路及 VLAN 分配信息。

```
Switch 4(config)#interface range GigabitEthernet 0/1-2      // 配置干道链路
Switch 4(config-if-range)#switchport mode trunk
Switch 4(config-if-range)#exit
Switch 4(config)#interface range GigabitEthernet 0/3-14     // 分配部门VLAN 30
Switch 4(config-if-range)#switchport access vlan 30
Switch 4(config-if-range)#exit
Switch 4(config)#interface range GigabitEthernet 0/15-24    // 分配部门VLAN 40
Switch 4(config-if-range)#switchport access vlan 40
Switch 4(config-if-range)#exit
```

（5）在三层核心交换机 Switch 1 上配置基于 SVI 多备份组的 VRRP。

① 配置 VLAN 10 的出口策略。

```
Switch 1(config)#interface VLAN 10
Switch 1(config-if-vlan 10)#vrrp 10 priority 120
```

/* 将交换机 Switch 1 划分到 VRRP 组 10 中，并配置较高优先级 120，从而成为 VRRP 组 10

多层交换技术(实践篇)

中的主路由器 */
```
Switch 1(config-if-vlan 10)#vrrp 10 ip 192.168.10.254
Switch 1(config-if-vlan 10)#vrrp 11 ip 192.168.10.253
                    // 将交换机 Switch 1 划分到 VRRP 组 11 中,优先级默认为 100
Switch 1(config-if-vlan 10)#exit
```
② 配置 VLAN 20 的出口策略。
```
Switch 1(config)#interface VLAN 20
Switch 1(config-if-vlan 20)#vrrp 20 ip 192.168.20.254
                    // 将交换机 Switch 1 划分到 VRRP 组 20 中
Switch 1(config-if-vlan 20)#vrrp 21 priority 120
                    // 将交换机 Switch 1 划分到 VRRP 组 21 中,并配置优先级为 120
Switch 1(config-if-vlan 20)#vrrp 21 ip 192.168.20.253
                    // 将交换机 Switch 1 划分到 VRRP 组 21 中
Switch 1(config-if-vlan 20)#exit
```
③ 配置 VLAN 30 的出口策略。
```
Switch 1(config)#interface VLAN 30
Switch 1(config-if-vlan 30)#vrrp 30 priority 120
                    // 将交换机 Switch 1 划分到 VRRP 组 30 中,配置优先级为 120
Switch 1(config-if-vlan 30)#vrrp 30 ip 192.168.30.254
Switch 1(config-if-vlan 30)#vrrp 31 ip 192.168.30.253
                    // 将交换机 Switch 1 划分到 VRRP 组 31 中
```
④ 配置 VLAN 40 的出口策略。
```
Switch 1(config)#interface VLAN 40
Switch 1(config-if-vlan 40)#vrrp 40 ip 192.168.40.254
                    // 将交换机 Switch 1 划分到 VRRP 组 40 中
Switch 1(config-if-vlan 40)#vrrp 41 priority 120
                    // 将交换机 Switch 1 划分到 VRRP 组 41 中,并配置优先级为 120
Switch 1(config-if-vlan 40)#vrrp 41 ip 192.168.40.253
Switch 1(config-if-vlan 40)#exit
```
(6) 在三层核心交换机 Switch 2 上配置基于 SVI 多备份组的 VRRP。
① 配置 VLAN 10 的出口策略。
```
Switch 2(config)#interface VLAN 10
Switch 2(config-if-vlan 10)#vrrp 10 ip 192.168.10.254
Switch 2(config-if-vlan 10)#vrrp 11 priority 120
Switch 2(config-if-vlan 10)#vrrp 11 ip 192.168.10.253
Switch 2(config-if-vlan 10)#exit
```
② 配置 VLAN 20 的出口策略。
```
Switch 2(config)#interface VLAN 20
Switch 2(config-if-vlan 20)#vrrp 20 priority 120
Switch 2(config-if-vlan 20)#vrrp 20 ip 192.168.20.254
```

任务 ⑫ 使用 VRRP 技术实现核心网络冗余备份（基于 SVI）

```
Switch 2(config-if-vlan 20)#vrrp 21 ip 192.168.20.253
Switch 2(config-if-vlan 20)#exit
```

③ 配置 VLAN 30 的出口策略。

```
Switch 2(config)#interface VLAN 30
Switch 2(config-if-vlan 30)#vrrp 30 ip 192.168.30.254
Switch 2(config-if-vlan 30)#vrrp 31 priority 120
Switch 2(config-if-vlan 30)#vrrp 31 ip 192.168.30.253
Switch 2(config-if-vlan 30)#exit
```

④ 配置 VLAN 40 的出口策略。

```
Switch 2(config)#interface VLAN 40
Switch 2(config-if-vlan 40)#vrrp 40 priority 120
Switch 2(config-if-vlan 40)#vrrp 40 ip 192.168.40.254
Switch 2(config-if-vlan 40)#vrrp 41 ip 192.168.40.253
Switch 2(config-if-vlan 40)#exit
```

【测试验证】

（1）在 Switch 1 上，使用"show vrrp brief"命令验证配置。

```
Switch 1#show vrrp brief
Interface    Grp  Pri  timer  Own  Pre  State   Master addr     Group addr
VLAN 10      10   120  9      -    P    Master  192.168.10.1    192.168.10.254
VLAN 10      11   100  9      -    P    Backup  192.168.10.2    192.168.10.253
VLAN 20      20   100  9      -    P    Backup  192.168.20.2    192.168.20.254
VLAN 20      21   120  9      -    P    Master  192.168.20.1    192.168.20.253
VLAN 30      30   120  9      -    P    Master  192.168.30.1    192.168.30.254
VLAN 30      31   100  9      -    P    Backup  192.168.30.2    192.168.30.253
VLAN 40      40   100  9      -    P    Backup  192.168.40.2    192.168.40.254
VLAN 40      41   120  9      -    P    Master  192.168.40.1    192.168.40.253
```

从 show 命令的输出结果可以看到，交换机 Switch 1 在 VRRP 组 10、21、30、41 中状态为主路由器，在 VRRP 组 11、20、31、40 中状态为备份路由器。

（2）在 Switch 2 上，使用"show vrrp brief"命令验证配置。

```
Switch 2#show vrrp brief
Interface    Grp  Pri  timer  Own  Pre  State   Master addr     Group addr
VLAN 10      10   100  9      -    P    Backup  192.168.10.1    192.168.10.254
VLAN 10      11   120  9      -    P    Master  192.168.10.2    192.168.10.253
VLAN 20      20   120  9      -    P    Master  192.168.20.2    192.168.20.254
VLAN 20      21   100  9      -    P    Backup  192.168.20.1    192.168.20.253
VLAN 30      30   100  9      -    P    Backup  192.168.30.1    192.168.30.254
VLAN 30      31   120  9      -    P    Master  192.168.30.2    192.168.30.253
VLAN 40      40   120  9      -    P    Master  192.168.40.2    192.168.40.254
VLAN 40      41   100  9      -    P    Backup  192.168.40.1    192.168.40.253
```

从 show 命令的输出结果可以看到，交换机 Switch 2 在 VRRP 组 10、21、30、41 中状

多层交换技术（实践篇）

态为备份路由器，在 VRRP 组 11、20、31、40 中状态为主路由器。

【注意事项】

（1）在 VRRP 多组中，需要为不同客户端配置不同的网关地址，以达到负载均衡的目的。

（2）在配置基于 SVI 的 VRRP 多备份组时，需要将 VRRP 组配置在 VLAN 端口下。

任务 ⑬ 使用 MSTP+VRRP+AP 技术保障核心网络稳健运行

【任务描述】

某企业为了提升网络稳健性，通过冗余和备份增强了网络的健壮性。此外，为了使核心网络以高带宽传输，通过聚合端口技术实现了逻辑链路的带宽聚合。针对核心层设备，希望通过 MSTP+VRRP 技术实现冗余网关，互相备份。

【任务目标】

掌握 MSTP、VRRP、AP 技术在核心网络中的综合应用。

【组网拓扑】

图 13-1 所示网络拓扑为某企业核心网络连接场景，通过实施 MSTP+VRRP+AP 综合技术，实现核心网络冗余、备份、高带宽、负载均衡。

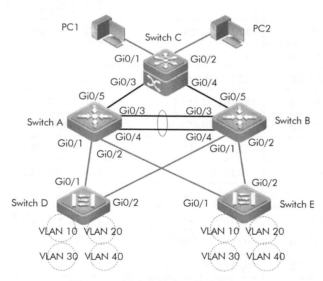

图 13-1 某企业核心网络连接场景

备注 1：配置前，先不要将交换机连接起来，配置完成后再进行连线，以免产生网络环路，造成网络故障。

备注 2：在拓扑图中，由于多条备份链路连接产生了环路，所以在开始时最好将某些端口阻塞（如汇聚交换机的 4 个端口在配置完毕测试时才打开），否则产生环路后，会发现设备的 CPU 利用率达到了 100%（可使用 "show cpu" 命令进行查看）。

备注 3：具体的设备连接接口信息，无论百兆口还是吉比特口，可根据具体情况决定。

多层交换技术（实践篇）

备注 4：全网设备的 IP 地址规划如表 13-1 所示。

表 13-1 全网设备的 IP 地址规划

设备名称	接口标识	接口地址	连接位置
Switch A	Gi0/1	\	连接 Switch D 的 Gi0/1
	Gi0/2	\	连接 Switch E 的 Gi0/1
	Gi0/3	\	连接 Switch B 的 Gi0/3
	Gi0/4	\	连接 Switch B 的 Gi0/4
	Gi0/5	172.168.1.1/24	连接 Switch C 的 Gi0/3
	VLAN 10	192.168.10.254/24	SVI 接口
	VLAN 20	192.168.20.253/24	SVI 接口
	VLAN 30	192.168.30.254/24	SVI 接口
	VLAN 40	192.168.40.253/24	SVI 接口
Switch B	Gi0/1	\	连接 Switch D 的 Gi0/2
	Gi0/2	\	连接 Switch E 的 Gi0/2
	Gi0/3	\	连接 Switch A 的 Gi0/3
	Gi0/4	\	连接 Switch A 的 Gi0/4
	Gi0/5	172.168.2.1/24	连接 Switch C 的 Gi0/4
	VLAN 10	192.168.10.253/24	SVI 接口
	VLAN 20	192.168.20.254/24	SVI 接口
	VLAN 30	192.168.30.253/24	SVI 接口
	VLAN 40	192.168.40.254/24	SVI 接口
Switch C	Gi0/1	172.168.6.1/24	连接测试 PC1
	Gi0/2	172.168.7.1/24	连接测试 PC2
	Gi0/3	172.168.1.2/24	连接 Switch A 的 Gi0/5
	Gi0/4	172.168.2.2/24	连接 Switch B 的 Gi0/5
VRRP	VLAN 10	192.168.10.250/24	VRRP 1
	VLAN 20	192.168.20.250/24	VRRP 2
	VLAN 30	192.168.30.250/24	VRRP 3
	VLAN 40	192.168.40.250/24	VRRP 4
Switch D	Gi0/1	\	连接 Switch A 的 Gi0/1
	Gi0/2	\	连接 Switch B 的 Gi0/1
Switch E	Gi0/1	\	连接 Switch A 的 Gi0/2
	Gi0/2	\	连接 Switch B 的 Gi0/2

任务 ⑬　使用 MSTP+VRRP+AP 技术保障核心网络稳健运行

【设备清单】

二层交换机（两台），三层交换机（3 台），网线（若干），PC（若干）。

【实施步骤】

（1）在接入交换机 Switch D 上配置设备基本信息。

① 在 Switch D 上创建 VLAN。

```
Ruijie#Configure terminal
Ruijie(config)#hostname Switch D
Switch D(config)#vlan 10              // 在接入交换机上创建 VLAN 10、20、30 和 40
Switch D(config-vlan)#exit
Switch D(config)#vlan 20
Switch D(config-vlan)#exit
Switch D(config)#vlan 30
Switch D(config-vlan)#exit
Switch D(config)#vlan 40
Switch D(config-vlan)#exit
```

② 在 Switch D 上配置接口信息。

```
Switch D(config)#
Switch D(config)#interface GigabitEthernet 0/1
Switch D(config-if-GigabitEthernet 0/1)#switchport mode trunk
                                    // 配置 Gi0/1 为 Trunk 端口
Switch D(config-if-GigabitEthernet 0/1)#exit
Switch D(config)#interface Gi0/2
Switch D(config-if-GigabitEthernet 0/2)#switchport mode trunk
Switch D(config-if-GigabitEthernet 0/2)#exit

Switch D(config)#interface range GigabitEthernet 0/3-10
Switch D(config-if-range)#switchport access vlan 10
                                    // 将接口划分到相关 VLAN 中
Switch D(config-if-range)#exit
Switch D(config)#interface range GigabitEthernet 0/11-15
Switch D(config-if-range)#switchport access vlan 20
Switch D(config-if-range)#exit
Switch D(config)#interface range GigabitEthernet 0/16-20
Switch D(config-if-range)#switchport access vlan 30
Switch D(config-if-range)#exit
Switch D(config)#interface range GigabitEthernet 0/21-24
Switch D(config-if-range)#switchport access vlan 40
Switch D(config-if-range)#exit
```

多层交换技术（实践篇）

（2）在接入交换机 Switch E 上配置设备基本信息。

① 在 Switch E 上创建 VLAN。

```
Ruijie#Configure terminal
Ruijie(config)#hostname Switch E
Switch E(config)#vlan 10              // 创建 VLAN10、20、30 和 40
Switch E(config-vlan)#exit
Switch E(config)#vlan 20
Switch E(config-vlan)#exit
Switch E(config)#vlan 30
Switch E(config-vlan)#exit
Switch E(config)#vlan 40
Switch E(config-vlan)#exit
```

② 在 Switch E 上配置接口信息。

```
Switch E(config)#interface Gi0/1                    // 配置 Gi0/1 和 Gi0/2 为 Trunk 端口
Switch E(config-if-GigabitEthernet 0/1)#switchport mode trunk
Switch E(config-if-GigabitEthernet 0/1)#exit
Switch E(config)#interface Gi0/2
Switch E(config-if-GigabitEthernet 0/2)#switchport mode trunk
Switch E(config-if-GigabitEthernet 0/2)#exit

Switch E(config)#interface range GigabitEthernet 0/3-10
                                     // 将接口划分到相关 VLAN 中
Switch E(config-if-range)#switchport access vlan 10
Switch E(config-if-range)#exit
Switch E(config)#interface range GigabitEthernet 0/11-15
Switch E(config-if-range)#switchport access vlan 20
Switch E(config-if-range)#exit
Switch E(config)#interface range GigabitEthernet 0/16-20
Switch E(config-if-range)#switchport access vlan 30
Switch E(config-if-range)#exit
Switch E(config)#interface range GigabitEthernet 0/21-24
Switch E(config-if-range)#switchport access vlan 40
Switch E(config-if-range)#exit
```

（3）在汇聚交换机 Switch A 上配置设备的 VLAN 信息。

需要在汇聚交换机 Switch A 上配置 VRRP，完成 VLAN 10、20、30、40 的虚拟 IP 地址、上行端口及 OSPF 动态路由配置。

① 在汇聚交换机 Switch A 上创建 SVI 接口。

```
Ruijie#Configure terminal
Ruijie(config)#hostname Switch A
Switch A(config)#vlan 10         // 创建 VLAN 10、20、30 和 40
```

任务 ⑬ 使用 MSTP+VRRP+AP 技术保障核心网络稳健运行

```
Switch A(config-vlan)#exit
Switch A(config)#vlan 20
Switch A(config-vlan)#exit
Switch A(config)#vlan 30
Switch A(config-vlan)#exit
Switch A(config)#vlan 40
Switch A(config-vlan)#exit
```

② 在汇聚交换机 Switch A 上配置 VLAN 10 的 SVI 接口 IP 地址及 VRRP 参数。

```
Switch A(config)#interface vlan 10
Switch A(config-if-vlan 10)#ip address 192.168.10.254 255.255.255.0
Switch A(config-if-vlan 10)#no shutdown    // 配置 VLAN 10 的 SVI 接口 IP 地址
/* 每一个 VRRP 组都有多台三层设备,在 Switch A 的 VLAN 10 上有这个地址,在 Switch B 的 VLAN 10 上也有相应同子网 IP 地址 */

Switch A(config-if-vlan 10)#vrrp 1 ip 192.168.10.250    // 配置虚拟 IP 地址
Switch A(config-if-vlan 10)#vrrp 1 preempt
/* 设为抢占模式。在正常状况下,VLAN 10 中的数据由 Switch A 传输。当 Switch A 发生故障时,由 Switch B 担负传输任务。若不配置抢占模式,当 Switch A 恢复正常后,仍由 Switch B 负责传输。配置抢占模式后,恢复正常的 Switch A 会再次夺取对 VLAN 10 的控制权 */
Switch A(config-if-vlan 10)#vrrp 1 priority 254
Switch A(config-if-vlan 10)#exit
```
/* 设备 VLAN 10 的 VRRP 优先级设为 254。在同一个 VLAN 中,优先级高的设备会成为 Master,优先级低的设备会成为 Backup */

③ 在汇聚交换机 Switch A 上配置 VLAN 20 的 SVI 接口 IP 地址及 VRRP 参数。

```
Switch A(config)#interface vlan 20
Switch A(config-if-vlan 20)#ip address 192.168.20.253 255.255.255.0
                                               // 配置 VLAN 20 的 IP 地址
Switch A(config-if-vlan 20)#no shutdown

Switch A(config-if-vlan 20)#vrrp 2 ip 192.168.20.250    // 配置虚拟 IP 地址
Switch A(config-if-vlan 20)#exit
```

④ 在汇聚交换机 Switch A 上配置 VLAN 30 的 SVI 接口 IP 地址及 VRRP 参数。

```
Switch A(config)#interface vlan 30
Switch A(config-if-vlan 30)#ip address 192.168.30.254 255.255.255.0
Switch A(config-if-vlan 30)#vrrp 3 ip 192.168.30.250
Switch A(config-if-vlan 30)#vrrp 3 preempt
Switch A(config-if-vlan 30)#vrrp 3 priority 254
Switch A(config-if-vlan 30)#exit
```

⑤ 在汇聚交换机 Switch A 上配置 VLAN 40 的 SVI 接口 IP 地址及 VRRP 参数。

多层交换技术（实践篇）

```
Switch A(config)#interface vlan 40
Switch A(config-if-vlan 40)#ip address 192.168.40.253 255.255.255.0
Switch A(config-if-vlan 40)#vrrp 4 ip 192.168.40.250
Switch A(config-if-vlan 40)#exit
```

⑥ 在汇聚交换机 Switch A 上配置接口信息。

```
Switch A(config)#interface GigabitEthernet 0/5
Switch A(config-if-GigabitEthernet 0/5)#no switchport
Switch A(config-if-GigabitEthernet 0/5)#ip address 172.16.1.1 255.255.255.0
Switch A(config-if-GigabitEthernet 0/5)#exit

Switch A(config)#interface GigabitEthernet 0/1                    // 配置Trunk端口
Switch A(config-if-GigabitEthernet 0/1)#switchport mode trunk
Switch A(config-if-GigabitEthernet 0/1)#interface GigabitEthernet 0/2
Switch A(config-if-GigabitEthernet 0/2)#switchport mode trunk
Switch A(config-if-GigabitEthernet 0/2)#interface GigabitEthernet 0/3
Switch A(config-if-GigabitEthernet 0/3)#switchport mode trunk
Switch A(config-if-GigabitEthernet 0/3)#interface GigabitEthernet 0/4
Switch A(config-if-GigabitEthernet 0/4)#switchport mode trunk
Switch A(config-if-GigabitEthernet 0/4)#end
```

⑦ 在汇聚交换机 Switch A 上配置全网路由信息。

```
Switch A(config)#router ospf                     // 配置OSPF动态路由
Switch A(config-router)#network 172.16.1.0 0.0.0.255 area 0
Switch A(config-router)#network 192.168.10.0 0.0.255.255 area 0
Switch A(config-router)#network 192.168.20.0 0.0.255.255 area 0
Switch A(config-router)#network 192.168.30.0 0.0.255.255 area 0
Switch A(config-router)#network 192.168.40.0 0.0.255.255 area 0
Switch A(config-router)#end
```

（4）在汇聚交换机 Switch B 上配置 Switch B 的 VLAN 信息。

需要在汇聚交换机 Switch B 上配置 VRRP，完成 VLAN 10、20、30、40 的虚拟 IP 地址、上行端口及 OSPF 动态路由配置。

① 在汇聚交换机 Switch B 上创建 SVI 接口。

```
Ruijie#Configure terminal
Ruijie(config)#hostname Switch B
Switch B(config)#vlan 10              // 创建 VLAN 10、20、30 和 40
Switch B(config-vlan)#exit
Switch B(config)#vlan 20
Switch B(config-vlan)#exit
Switch B(config)#vlan 30
Switch B(config-vlan)#exit
```

任务 ⑬ 使用 MSTP+VRRP+AP 技术保障核心网络稳健运行

```
Switch B(config)#vlan 40
Switch B(config-vlan)#exit
```
② 在汇聚交换机 Switch B 上配置 VLAN 10 的 SVI 接口 IP 地址及 VRRP 参数。
```
Switch B(config)#interface vlan 10
Switch B(config-if-vlan 10)#ip address 192.168.10.253 255.255.255.0
Switch B(config-if-vlan 10)#no shutdown
Switch B(config-if-vlan 10)#vrrp 1 ip 192.168.10.250
                          // VLAN 10 的 VRRP 不设置优先级，默认为 100
Switch B(config-if-vlan 10)#exit
```
③ 在汇聚交换机 Switch B 上配置 VLAN 20 的 SVI 接口 IP 地址及 VRRP 参数。
```
Switch B(config)#interface vlan 20
Switch B(config-if-vlan 20)#ip address 192.168.30.254 255.255.255.0
Switch B(config-if-vlan 20)#vrrp 2 ip 192.168.20.250
Switch B(config-if-vlan 20)#vrrp 2 preempt
Switch B(config-if-vlan 20)#vrrp 2 priority 254
Switch B(config-if-vlan 20)#exit
```
④ 在汇聚交换机 Switch B 上配置 VLAN 30 的 SVI 接口 IP 地址及 VRRP 参数。
```
Switch B(config)#interface vlan 30
Switch B(config-if-vlan 30)#ip address 192.168.30.253 255.255.255.0
Switch B(config-if-vlan 30)#vrrp 3 ip 192.168.30.250
Switch B(config-if-vlan 30)#exit
```
⑤ 在汇聚交换机 Switch B 上配置 VLAN 40 的 SVI 接口 IP 地址及 VRRP 参数。
```
Switch B(config)#interface vlan 40
Switch B(config-if-vlan 40)#ip address 192.168.40.254 255.255.255.0
Switch B(config-if-vlan 40)#vrrp 4 ip 192.168.40.250
Switch B(config-if-vlan 40)#vrrp 4 preempt
Switch B(config-if-vlan 40)#vrrp 4 priority 254
Switch B(config-if-vlan 40)#exit
```
⑥ 在汇聚交换机 Switch B 上配置接口信息。
```
Switch B(config)#interface GigabitEthernet 0/5
Switch B(config-if-GigabitEthernet 0/5)#no switchport
Switch B(config-if-GigabitEthernet 0/5)#ip address 172.16.2.1 255.255.255.0
Switch B(config-if-GigabitEthernet 0/1)#exit

Switch B(config)#interface GigabitEthernet0/1        // 配置连接的 Trunk 端口
Switch B(config-if-GigabitEthernet 0/1)#switchport mode trunk
Switch B(config-if-GigabitEthernet 0/1)#interface GigabitEthernet 0/2
Switch B(config-if-GigabitEthernet 0/2)#switchport mode trunk
Switch B(config-if-GigabitEthernet 0/2)#interface GigabitEthernet 0/3
Switch B(config-if-GigabitEthernet 0/3)#switchport mode trunk
```

多层交换技术（实践篇）

```
Switch B(config-if-GigabitEthernet 0/3)#interface GigabitEthernet 0/4
Switch B(config-if-GigabitEthernet 0/4)#switchport mode trunk
Switch B(config-if-GigabitEthernet 0/4)#end
```

⑦ 在三层汇聚交换机 Switch B 上配置全网路由信息。

```
Switch B(config)#router ospf                    // 配置OSPF动态路由
Switch B(config-router)#network 192.168.10.0 0.0.255.255 area 0
Switch B(config-router)#network 192.168.20.0 0.0.255.255 area 0
Switch B(config-router)#network 192.168.30.0 0.0.255.255 area 0
Switch B(config-router)#network 192.168.40.0 0.0.255.255 area 0
Switch B(config-router)#network 172.16.2.0 0.0.0.255 area 0
Switch B(config-router)#end
```

（5）在三层汇聚交换机 Switch C 上配置 Switch C 基本信息。

Switch C 作为核心层要执行转发功能，配置上下行端口 IP 地址及 OSPF 动态路由。

① 在三层汇聚交换机 Switch C 上配置接口地址，完成转发功能。

```
Ruijie#Configure terminal
Ruijie(config)#hostname Switch C
Switch C(config)#interface GigabitEthernet 0/1
Switch C(config-if-GigabitEthernet 0/1)#no switchport
Switch C(config-if-GigabitEthernet 0/1)#ip address 172.16.6.1 255.255.255.0
Switch C(config-if-GigabitEthernet 0/1)#exit
Switch C(config)#interface GigabitEthernet 0/2
Switch C(config-if-GigabitEthernet 0/2)#no switchport
Switch C(config-if-GigabitEthernet 0/2)#ip address 172.16.7.1 255.255.255.0
Switch C(config-if-GigabitEthernet 0/2)#exit

Switch C(config)#interface GigabitEthernet 0/3
Switch C(config-if-GigabitEthernet 0/3)#no switchport
Switch C(config-if-GigabitEthernet 0/3)#ip address 172.16.1.2 255.255.255.0
Switch C(config-if-GigabitEthernet 0/3)#exit
Switch C(config)#interface GigabitEthernet 0/4
Switch C(config-if-GigabitEthernet 0/4)#no switchport
Switch C(config-if-GigabitEthernet 0/4)#ip address 172.16.2.2 255.255.255.0
Switch C(config-if-GigabitEthernet 0/4)#exit
```

② 在三层汇聚交换机 Switch C 上配置 OSPF 动态路由，实现全网互通。

```
Switch C(config)#router ospf                    // 配置OSPF动态路由
Switch C(config-router)#network 172.16.1.0 0.0.0.255 area 0
Switch C(config-router)#network 172.16.2.0 0.0.0.255 area 0
Switch C(config-router)#network 172.16.6.0 0.0.0.255 area 0
```

任务 ⑬ 使用 MSTP+VRRP+AP 技术保障核心网络稳健运行

```
Switch C(config-router)#network 172.16.7.0 0.0.0.255 area 0
Switch C(config-router)#end
```

（6）配置三层汇聚交换机 Switch A 与 Switch B 的骨干链路端口聚合。

配置 Switch A 和 Switch B 的骨干链路 Gi0/3 和 Gi0/4 端口为 Trunk 端口，以获得高带宽。此外，一定要将 AggregatePort 的 Switchport Mode 配置为 Trunk 模式，否则其默认为 Access 模式。

```
Switch A(config)#interface GigabitEthernet 0/3
Switch A(config-if-GigabitEthernet 0/3)#port-group 1    // 加入端口聚合1组
Switch A(config-if-GigabitEthernet 0/3)#exit
Switch A(config)#interface GigabitEthernet 0/4
Switch A(config-if-GigabitEthernet 0/4)#port-group 1
Switch A(config-if-GigabitEthernet 0/4)#end

Switch A(config)#interface aggregatePort 1
Switch A(config-if-aggregatePort 1)#switchport mode trunk
Switch A(config-if-aggregatePort 1)#exit

Switch B(config)#
Switch B(config)#interface GigabitEthernet 0/3
Switch B(config-if-GigabitEthernet 0/3)#port-group 1    // 加入端口聚合1组
Switch B(config-if-GigabitEthernet 0/3)#exit
Switch B(config)#interface GigabitEthernet 0/4
Switch B(config-if-GigabitEthernet 0/4)#port-group 1
Switch B(config-if-GigabitEthernet 0/4)#end

Switch B(config)#interface aggregatePort 1
Switch B(config-if-aggregatePort 1)#switchport mode trunk
```

（7）查看 Switch A 与 Switch B 的端口聚合状态。

```
Switch A#show aggregatePort 1 summary
AggregatePort    MaxPorts    SwitchPort    Mode     Ports
-------------    --------    ----------    ----     -----------------
Ag1              8           Enabled       Trunk    Fa0/3 , Fa0/4


Switch B#show aggregatePort 1 summary
AggregatePort    MaxPorts    SwitchPort    Mode     Ports
-------------    --------    ----------    ----     -----------------
Ag1              8           Enabled       Trunk    Fa0/3 , Fa0/4


Switch A#show ip route              // 查看路由表
Type: C - connected, S - static, R - RIP, O - OSPF, IA - OSPF inter area
      N1 - OSPF NSSA external type 1, N2 - OSPF NSSA external type 2
```

多层交换技术（实践篇）

```
        E1 - OSPF external type 1, E2 - OSPF external type 2
Type  Destination IP     Next hop       Interface  Distance  Metric  Status
---   --------------     ----------     ---------  --------  ------  ------
C     172.16.1.0/24      0.0.0.0        Fa0/5      0         0       Active
O     172.16.2.0/24      172.16.1.2     Fa0/5      110       2       Active
O     172.16.6.0/24      172.16.1.2     Fa0/5      110       2       Active
O     172.16.7.0/24      172.16.1.2     Fa0/5      110       2       Active
C     192.168.10.0/24    0.0.0.0        VL10       0         0       Active
C     192.168.20.0/24    0.0.0.0        VL20       0         0       Active
C     192.168.30.0/24    0.0.0.0        VL30       0         0       Active
C     192.168.40.0/24    0.0.0.0        VL40       0         0       Active

Switch A#show vrrp                     // 查看 VRRP 应用情况
If    Group  State    Priority  Preempt  Interval  Virtual IP        Auth
----  -----  ------   --------  -------  --------  ---------------   ----
VL10  1      master   254       may      1         192.168.10.250
VL20  2      backup   100       may      1         192.168.20.250
VL30  3      master   254       may      1         192.168.30.250
VL40  4      backup   100       may      1         192.168.40.250

Switch B#show ip route
Type: C - connected, S - static, R - RIP, O - OSPF, IA - OSPF inter area
      N1 - OSPF NSSA external type 1, N2 - OSPF NSSA external type 2
      E1 - OSPF external type 1, E2 - OSPF external type 2

Type  Destination IP     Next hop       Interface  Distance  Metric  Status
---   --------------     ----------     ---------  --------  ------  --------
O     172.16.1.0/24      172.16.2.2     Fa0/5      110       2       Active
C     172.16.2.0/24      0.0.0.0        Fa0/5      0         0       Active
O     172.16.6.0/24      172.16.2.2     Fa0/5      110       2       Active
O     172.16.7.0/24      172.16.2.2     Fa0/5      110       2       Active
C     192.168.10.0/24    0.0.0.0        VL10       0         0       Inactive
O     192.168.10.0/24    172.16.2.2     Fa0/5      110       3       Active
C     192.168.20.0/24    0.0.0.0        VL20       0         0       Inactive
O     192.168.20.0/24    172.16.2.2     Fa0/5      110       3       Active
C     192.168.30.0/24    0.0.0.0        VL30       0         0       Inactive
O     192.168.30.0/24    172.16.2.2     Fa0/5      110       3       Active
C     192.168.40.0/24    0.0.0.0        VL40       0         0       Inactive
O     192.168.40.0/24    172.16.2.2     Fa0/5      110       3       Active

/* 在这里可以看到，汇聚交换机上有 4 条 Inactive 路由，这是因为 Gi0/1 和 Gi0/2 接口没有
```

任务 ⑬ 使用 MSTP+VRRP+AP 技术保障核心网络稳健运行

开启，本身的 VLAN 没有生效 */

```
Switch B#show vrrp           // 查看 VRRP 应用情况
If    Group State    Priority Preempt Interval Virtual IP      Auth
--  - --  ----      ----     ------- -------- ----------      ----
VL10  1    backup   100      may     1        192.168.10.250
VL20  2    master   254      may     1        192.168.20.250
VL30  3    backup   100      may     1        192.168.30.250
VL40  4    master   254      may     1        192.168.40.250

Switch C#show ip route                           // 查看路由表
Type:  C - connected,  S - static,  R - RIP,  O - OSPF,  IA - OSPF inter area
       N1 - OSPF NSSA external type 1,  N2 - OSPF NSSA external type 2
       E1 - OSPF external type 1,  E2 - OSPF external type 2

Type Destination IP    Next hop      Interface Distance Metric  Status
---  --------------    ----------    --------- -------- ------  ------
C    172.16.1.0/24     0.0.0.0       Fa0/3     0        0       Active
C    172.16.2.0/24     0.0.0.0       Fa0/4     0        0       Active
C    172.16.6.0/24     0.0.0.0       Fa0/1     0        0       Active
C    172.16.7.0/24     0.0.0.0       Fa0/2     0        0       Active
O    192.168.10.0/24   172.16.1.1    Fa0/3     110      2       Active
O    192.168.20.0/24   172.16.1.1    Fa0/3     110      2       Active
O    192.168.30.0/24   172.16.1.1    Fa0/3     110      2       Active
O    192.168.40.0/24   172.16.1.1    Fa0/3     110      2       Active
```

VRRP 配置完成后，从二层接入交换机上发送到三层汇聚交换机 Switch D、Switch E 上的报文通过三层汇聚交换机转发。其中，VLAN 10 和 VLAN 30 中的数据通过 Switch A 上行；VLAN20 和 VLAN 40 中的数据通过 Switch B 上行。通过全网中实施的 MSTP+VRRP 技术，保障网络健壮性的同时，还实现了网络中不同 VLAN 通信流量的负载均衡。

但是，来自外部网络的数据包通过 Switch C 向内部网络传输时，纯粹按照 Switch C 上的路由表进行发送，因此在数据上行时可以实现负载均衡，但是下行时无法实现负载均衡，所以需要进行路由优先级的配置。

（8）配置 Switch A、Switch B 的路由优先级。

```
Switch A(config)#
Switch A(config)#interface vlan 20
Switch A(config-if-vlan 20)#ip ospf cost 65535
                                // 将 VLAN 20 的 OSPF 开销值设为 65535
Switch A(config-if-vlan 20)#exit
Switch A(config)#interface vlan 40
Switch A(config-if-vlan 40)#ip ospf cost 65535
```

```
                                           // 将 VLAN 40 的 OSPF 开销值设为 65535
Switch A(config-if-vlan 40)#exit

Switch B(config)#
Switch B(config)#interface vlan 10
Switch B(config-if-vlan 10)#ip ospf cost 65535
                                           // 将 VLAN 10 的 OSPF 开销值设为 65535
Switch B(config-if-vlan 10)#exit
Switch B(config)#interface vlan 30
Switch B(config-if-vlan 30)#ip ospf cost 65535
                                           // 将 VLAN 30 的 OSPF 开销值设为 65535
Switch B(config-if-vlan 30)#exit
```

（9）查看核心设备 Switch C 的路由表。

```
Switch C#show ip route        // 查看路由表信息
Type: C - connected,  S - static,  R - RIP,  O - OSPF,  IA - OSPF inter area
      N1 - OSPF NSSA external type 1,  N2 - OSPF NSSA external type 2
      E1 - OSPF external type 1,  E2 - OSPF external type 2

Type  Destination IP      Next hop      Interface  Distance  Metric   Status
----  ----------------    ----------    ---------  --------  ------   ------
C     172.16.1.0/24       0.0.0.0       Fa0/3      0         0        Active
C     172.16.2.0/24       0.0.0.0       Fa0/4      0         0        Active
C     172.16.6.0/24       0.0.0.0       Fa0/1      0         0        Active
C     172.16.7.0/24       0.0.0.0       Fa0/2      0         0        Active
O     192.168.10.0/24     172.16.1.1    Fa0/3      110       2        Active
O     192.168.20.0/24     172.16.1.1    Fa0/3      110       65536    Active
O     192.168.30.0/24     172.16.1.1    Fa0/3      110       2        Active
O     192.168.40.0/24     172.16.1.1    Fa0/3      110       65536    Active
```

和之前的路由表对比，指向 Switch A 上 VLAN 20 和 40 的路由 Metric 值发生了变化。由于 Switch B 的 Gi0/1 和 Gi0/2 口没有打开，VLAN 的地址没有生效，所以即使 Switch A 的路由开销值达到了最大，也仍会由 Switch A 转发。

至此，VRRP 配置完毕。由于 Switch B 的 Gi0/1 和 Gi0/2 口没有打开，所以还无法使用，需要配置 MSTP。

（10）配置接入交换机 Switch D 的 MSTP，并检查生成树信息。

在 Switch D 上开启生成树，状态设为 MSTP，并配置实例和版本。

```
Switch D(config)#
Switch D(config)#spanning-tree                              // 开启生成树
Switch D(config)#spanning-tree mode mstp                    // 生成树类型为多生成树
Switch D(config)#spanning-tree mst configuration            // 配置多生成树
```

任务 ⑬　使用 MSTP+VRRP+AP 技术保障核心网络稳健运行

```
Switch D(config-mst)#instance 1 VLAN 10, 30
/* 将 VLAN 10、VLAN 30 放入实例 1。一个实例生成一个树，该生成树可以和其他实例生成的树的路径不一样，以实现负载均衡 */

Switch D(config-mst)#revision 1                    // 配置多生成树的版本号
Switch D(config-mst)#instance 2 vlan 20, 40
                                                   // 将 VLAN 20、VLAN 40 放入实例 2
Switch D(config-mst)#revision 1
Switch D(config-mst)#exit

Switch D# show spanning-tree                       // 查看生成树信息
StpVersion : MSTP
SysStpStatus : Enabled
BaseNumPorts : 24
MaxAge : 20
HelloTime : 2
ForwardDelay : 15
BridgeMaxAge : 20
BridgeHelloTime : 2
BridgeForwardDelay : 15
MaxHops : 20
TxHoldCount : 3
PathCostMethod : Long
BPDUGuard : Disabled
BPDUFilter : Disabled

###### MST 0 vlans mapped : 1-9, 11-19, 21-29,
                            31-39, 41-4094
BridgeAddr : 00d0.f8db.a401
Priority : 32768
TimeSinceTopologyChange : 0d:0h:15m:0s
TopologyChanges : 0
DesignatedRoot : 800000D0F8DBA401
RootCost : 0
RootPort : 0
CistRegionRoot : 800000D0F8DBA401
CistPathCost : 0

###### MST 1 vlans mapped : 10, 30
BridgeAddr : 00d0.f8db.a401
```

```
Priority : 32768
TimeSinceTopologyChange : 0d:18h:49m:57s
TopologyChanges : 0
DesignatedRoot : 800100D0F8DBA401
RootCost : 0
RootPort : 0

###### MST 2 vlans mapped : 20, 40
BridgeAddr : 00d0.f8db.a401
Priority : 32768
TimeSinceTopologyChange : 0d:18h:49m:57s
TopologyChanges : 0
DesignatedRoot : 800200D0F8DBA401
RootCost : 0
RootPort : 0

Switch D#show spanning-tree mst configuration    // 查看多生成树的配置信息
Multi spanning tree protocol : Enabled
Name        :
Revision : 1
Instance    Vlans Mapped
--------    ----------------------------------------------------------------
0           1-9, 11-19, 21-29,
            31-39, 41-4094
1           10, 30
2           20, 40
----------------------------------------------------------------------------
```

（11）配置接入交换机 Switch E 的 MSTP，并检查生成树信息。

在 Switch E 上开启生成树，状态设为 MSTP，配置实例和版本。

```
Switch E(config)#
Switch E(config)#spanning-tree                          // 开启生成树
Switch E(config)#spanning-tree mode mstp                // 生成树类型为多生成树
Switch E(config)#spanning-tree mst configuration        // 配置多生成树
Switch E(config-mst)#instance 1 vlan 10, 30
                                // 将 VLAN 10、VLAN 30 放入实例 1

Switch E(config-mst)#revision 1                         // 配置多生成树的版本号
Switch E(config-mst)#instance 2 vlan 20, 40
                                // 将 VLAN 20、VLAN 40 放入实例 2
Switch E(config-mst)#revision 1
```

任务 ⑬ 使用 MSTP+VRRP+AP 技术保障核心网络稳健运行

```
Switch E(config-mst)#exit


Switch E#show spanning-tree                        // 查看生成树信息
…
Switch E#show spanning-tree mst configuration      // 查看多生成树的配置信息
…
/* 查看生成树的配置信息，上述两项结果和 Switch D 的配置基本类似，只是 MAC 地址不一样，
限于篇幅，此处省略 */
```

（12）配置汇聚交换机 Switch A 的 MSTP 信息，并检查生成树消息。

在 Switch A 上开启生成树，状态设为 MSTP，配置实例和版本，并配置实例的优先级。

```
Switch A(config)#spanning-tree
Switch A(config)#spanning-tree mode mstp           // 生成树类型为多生成树
Switch A(config)#spanning-tree mst configuration   // 配置多生成树
Switch A(config-mst)#instance 1 vlan 10, 30
                                                   // 将 VLAN 10、VLAN 30 放入实例 1
Switch A(config-mst)#revision 1

Switch A(config-mst)#instance 2 vlan 20, 40
                                                   // 将 VLAN 20、VLAN 40 放入实例 2
Switch A(config-mst)#revision 1
Switch A(config-mst)#exit

Switch A(config)#spanning-tree mst 1 priority 4096
                                 // 配置实例 1 在 Switch A 上的优先级为 4096
Switch A(config)#spanning-tree mst 2 priority 8192
                                 // 配置实例 2 在 Switch A 上的先级为 8192
/* 配置优先级比较高是为了使 Switch A 成为 MST 1 的根节点。这一方面是因为它的性能比接
入交换机 Switch D 好，可以防止接入交换机被选作根节点。另一方面，如果默认优先级更高的为
Switch B，则 VLAN 10、VLAN 30 也会通过 Switch B 传输，这与希望的配置产生了冲突。因此，
要为 Switch B 配置次高优先级 */


Switch A#show spanning-tree                        // 查看生成树信息
StpVersion : MSTP
SysStpStatus : Enabled
BaseNumPorts : 24
MaxAge : 20
HelloTime : 2
ForwardDelay : 15
BridgeMaxAge : 20
BridgeHelloTime : 2
```

多层交换技术（实践篇）

```
BridgeForwardDelay : 15
MaxHops : 20
TxHoldCount : 3
PathCostMethod : Long
BPDUGuard : Disabled
BPDUFilter : Disabled

###### MST 0 vlans mapped : 1-9, 11-19, 21-29, 31-39, 41-4094
BridgeAddr : 00d0.f8ff.525f
Priority : 32768
TimeSinceTopologyChange : 0d:0h:10m:38s
TopologyChanges : 0
DesignatedRoot : 800000D0F8BFFE71
RootCost : 0
RootPort : Fa0/2
CistRegionRoot : 800000D0F8BFFE71
CistPathCost : 200000

###### MST 1 vlans mapped : 10, 30
BridgeAddr : 00d0.f8ff.525f
Priority : 4096
TimeSinceTopologyChange : 0d:19h:44m:27s
TopologyChanges : 0
DesignatedRoot : 100100D0F8FF525F
RootCost : 0
RootPort : 0

###### MST 2 vlans mapped : 20, 40
BridgeAddr : 00d0.f8ff.525f
Priority : 8192
TimeSinceTopologyChange : 0d:19h:44m:27s
TopologyChanges : 0
DesignatedRoot : 200200D0F8FF525F
RootCost : 0
RootPort : 0

Switch A#show spanning-tree mst configuration          // 查看生成树的配置信息
Multi spanning tree protocol : Enabled
Name :
Revision : 1
```

任务 ⓭ 使用 MSTP+VRRP+AP 技术保障核心网络稳健运行

```
Instance   Vlans Mapped
--------   ---------------------------------------------------
0          1-9, 11-19, 21-29, 31-39, 41-4094
1          10, 30
2          20, 40
--------   ---------------------------------------------------
```

（13）配置汇聚交换机 Switch B 的 MSTP 信息，并检查生成树消息。

在 Switch B 上开启生成树，状态设为 MSTP，配置实例和版本，并配置实例的优先级。

```
Switch B(config)#
Switch B(config)#spanning-tree
Switch B(config)#spanning-tree mode mstp         // 生成树类型为多生成树
Switch B(config)#spanning-tree mst configuration
Switch B(config-mst)#instance 1 vlan 10, 30
                                                 // 将 VLAN 10、VLAN 30 放入实例 1
Switch B(config-mst)#revision 1
Switch B(config-mst)#instance 2 vlan 20, 40
                                                 // 将 VLAN 20、VLAN 40 放入实例 2
Switch B(config-mst)#revision 1
Switch B(config-mst)#exit

Switch B(config)#spanning-tree mst 2 priority 4096   // 配置实例 2 优先级为 4096
Switch B(config)#spanning-tree mst 1 priority 8192   // 配置实例 1 优先级为 8192
Switch B(config)#interface range GigabitEthernet 0/1-2
// MSTP 设置已完成，可以打开初始时关闭的 Gi0/1 和 Gi0/2
Switch B(config-if-range)#no shutdown
```

```
Switch B#show spanning-tree                          // 查看生成树信息
StpVersion : MSTP
SysStpStatus : Enabled
BaseNumPorts : 24
MaxAge : 20
HelloTime : 2
ForwardDelay : 15
BridgeMaxAge : 20
BridgeHelloTime : 2
BridgeForwardDelay : 15
MaxHops : 20
TxHoldCount : 3
PathCostMethod : Long
```

多层交换技术（实践篇）

```
        BPDUGuard : Disabled
        BPDUFilter : Disabled

        ###### MST 0 vlans mapped : 1-9, 11-19, 21-29, 31-39, 41-4094
        BridgeAddr : 00d0.f8b8.3287
        Priority : 32768
        TimeSinceTopologyChange : 0d:0h:3m:23s
        TopologyChanges : 0
        DesignatedRoot : 800000D0F8B83287
        RootCost : 0
        RootPort : 0
        CistRegionRoot : 800000D0F8B83287
        CistPathCost : 0

        ###### MST 1 vlans mapped : 10, 30
        BridgeAddr : 00d0.f8b8.3287
        Priority : 8192
        TimeSinceTopologyChange : 0d:19h:55m:25s
        TopologyChanges : 0
        DesignatedRoot : 100100D0F8FF525F
        RootCost : 190000
        RootPort : Ag1

        ###### MST 2 vlans mapped : 20, 40
        BridgeAddr : 00d0.f8b8.3287
        Priority : 4096
        TimeSinceTopologyChange : 0d:19h:55m:25s
        TopologyChanges : 0
        DesignatedRoot : 100200D0F8B83287
        RootCost : 0
        RootPort : 0

        Switch B#show spanning-tree mst configuration            // 查看生成树的配置信息
        Multi spanning tree protocol : Enabled
        Name     :
        Revision : 1
        Instance   Vlans Mapped
        --------   ------------------------------------------------------------
        0          1-9, 11-19, 21-29, 31-39, 41-4094
```

任务 ⑬ 使用 MSTP+VRRP+AP 技术保障核心网络稳健运行

```
1          10, 30
2          20, 40
-------------------------------------------------------------------
```

（14）在汇聚交换机上查看配置完成的 VRRP 信息。

```
Switch A#show vrrp
If    Group State   Priority Preempt Interval Virtual IP       Auth
--  - --  -----  ----  -------- ------- -------- ------------------
VL10  1    master  254       may     1        192.168.10.250
VL20  2    backup  100       may     1        192.168.20.250
VL30  3    master  254       may     1        192.168.30.250
VL40  4    backup  100       may     1        192.168.40.250


Switch A#show ip route
Type: C - connected, S - static, R - RIP, O - OSPF, IA - OSPF inter area
      N1 - OSPF NSSA external type 1, N2 - OSPF NSSA external type 2
      E1 - OSPF external type 1, E2 - OSPF external type 2


Type Destination IP      Next hop       Interface Distance Metric   Status
---- ---------------     ----------     --------- -------- ------   ------
C    172.16.1.0/24       0.0.0.0        Fa0/5     0        0        Active
O    172.16.2.0/24       172.16.1.2     Fa0/5     110      2        Active
O    172.16.6.0/24       172.16.1.2     Fa0/5     110      2        Active
O    172.16.7.0/24       172.16.1.2     Fa0/5     110      2        Active
C    192.168.10.0/24     0.0.0.0        VL10      0        0        Active
C    192.168.20.0/24     0.0.0.0        VL20      0        0        Active
O    192.168.20.0/24     172.16.1.2     Fa0/5     110      3        Active
C    192.168.30.0/24     0.0.0.0        VL30      0        0        Active
C    192.168.40.0/24     0.0.0.0        VL40      0        0        Active
O    192.168.40.0/24     172.16.1.2     Fa0/5     110      3        Active
```

其中，学习到两条路由 192.168.20.0/24 和 192.168.40.0/24，其中，第一条是通过直连路由学习到的。同时，Switch B 上也存在这两个 IP 地址，所以通过 Switch C 上的 OSPF 路由学习到了 Switch B 的 192.168.20.0/24 和 192.168.40.0/24 两条路由。

```
Switch B#show vrrp
If    Group State   Priority Preempt Interval Virtual IP       Auth
--  - --  -----  ----  -------- ------- -------- ------------------
VL10  1    master  254       may     1        192.168.10.250
VL20  2    backup  100       may     1        192.168.20.250
VL30  3    master  254       may     1        192.168.30.250
VL40  4    backup  100       may     1        192.168.40.250
```

多层交换技术（实践篇）

```
Switch B#show ip route
Type: C - connected, S - static, R - RIP, O - OSPF, IA - OSPF inter area
      N1 - OSPF NSSA external type 1, N2 - OSPF NSSA external type 2
      E1 - OSPF external type 1, E2 - OSPF external type 2

Type  Destination IP      Next hop        Interface  Distance  Metric  Status
---   ---------------     ----------      ---------  --------  ------  ------
O     172.16.1.0/24       172.16.2.2      Fa0/5      110       2       Active
C     172.16.2.0/24       0.0.0.0         Fa0/5      0         0       Active
O     172.16.6.0/24       172.16.2.2      Fa0/5      110       2       Active
O     172.16.7.0/24       172.16.2.2      Fa0/5      110       2       Active
C     192.168.10.0/24     0.0.0.0         VL10       0         0       Active
O     192.168.10.0/24     172.16.2.2      Fa0/5      110       3       Active
C     192.168.20.0/24     0.0.0.0         VL20       0         0       Active
C     192.168.30.0/24     0.0.0.0         VL30       0         0       Active
O     192.168.30.0/24     172.16.2.2      Fa0/5      110       3       Active
C     192.168.40.0/24     0.0.0.0         VL40       0         0       Active
```

其中，学习到两条路由 192.168.10.0/24 和 192.168.30.0/24，第一条是通过直连路由学习到的。同时，核心设备上也存在这两个 IP 地址，所以通过 Switch C 上的 OSPF 路由学习到了 Switch A 的 192.168.10.0/24 和 192.168.30.0/24 两条路由。

```
Switch C#show ip route
Type: C - connected, S - static, R - RIP, O - OSPF, IA - OSPF inter area
      N1 - OSPF NSSA external type 1, N2 - OSPF NSSA external type 2
      E1 - OSPF external type 1, E2 - OSPF external type 2

Type  Destination IP      Next hop        Interface  Distance  Metric  Status
---   ---------------     ----------      ---------  --------  ------  ------
C     172.16.1.0/24       0.0.0.0         Fa0/3      0         0       Active
C     172.16.2.0/24       0.0.0.0         Fa0/4      0         0       Active
C     172.16.6.0/24       0.0.0.0         Fa0/1      0         0       Active
C     172.16.7.0/24       0.0.0.0         Fa0/2      0         0       Active
O     192.168.10.0/24     172.16.1.1      Fa0/3      110       2       Active
O     192.168.20.0/24     172.16.2.1      Fa0/4      110       2       Active
O     192.168.30.0/24     172.16.1.1      Fa0/3      110       2       Active
O     192.168.40.0/24     172.16.2.1      Fa0/4      110       2       Active
```

通过对比 Switch C 上的路由表可以发现，路由优先级的配置生效了。其中：VLAN 10 和 VLAN 30 中的数据通过 Switch A 下行；VLAN 20 和 VLAN 40 中的数据通过 Switch B 下行。至此，完成了"VRRP 由 MSTP+AP 端口聚合"配置。

（15）在三层汇聚交换机上测试验证网络连通性。

任务 ⑬ 使用 MSTP+VRRP+AP 技术保障核心网络稳健运行

```
Switch C#ping 192.168.10.254
Sending 5, 100-byte ICMP Echos to 192.168.10.254,
timeout is 2000 milliseconds.
!!!!!!!
Success rate is 100 percent (5/5)
Minimum = 1ms Maximum = 1ms, Average = 1ms

SWITCH C#ping 192.168.10.253
Sending 5, 100-byte ICMP Echos to 192.168.10.253,
timeout is 2000 milliseconds.
...
Success rate is 0 percent (0/5)
```

在 Switch C 上可以使用 Ping 命令进行测试，能 Ping 通 Switch A 上所连的 VLAN 10 中的设备，却 Ping 不通 Switch B 上所连的 VLAN 10 中的设备，可见 VLAN 10 中设备上的数据都通过 Switch A 传输。VLAN 20、VLAN 30、VLAN 40 具有相同的结果。

在 VLAN 10 的接口下，对测试 PC 做相关设置，如图 13-2 所示。

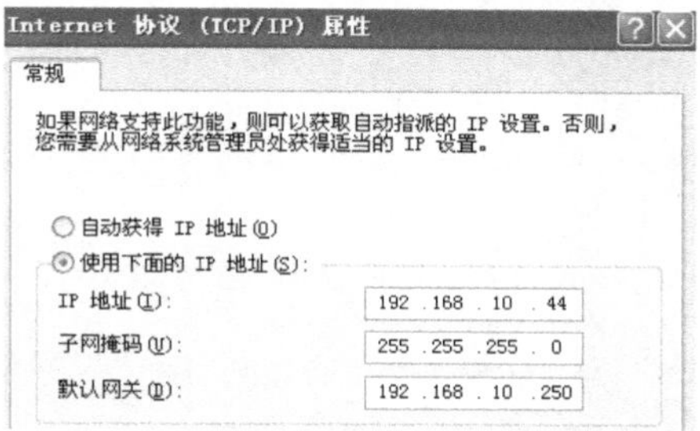

图 13-2 测试 PC 的相关设置（1）

在系统中运行 "C:\>tracert 172.16.6.1" 命令，结果如下。

```
Tracing route to 172.16.6.1 over a maximum of 30 hops
  1    1 ms    <1 ms    <1 ms    192.168.10.254      // VLAN 10 的 IP 地址
  2    <1 ms   <1 ms    <1 ms    172.16.6.1
Trace complete.
```

可见，VLAN 10 的数据是通过 Switch A 传输的。

同样，在 VLAN 20 的接口下，对测试 PC 做相关设置，如图 13-3 所示。

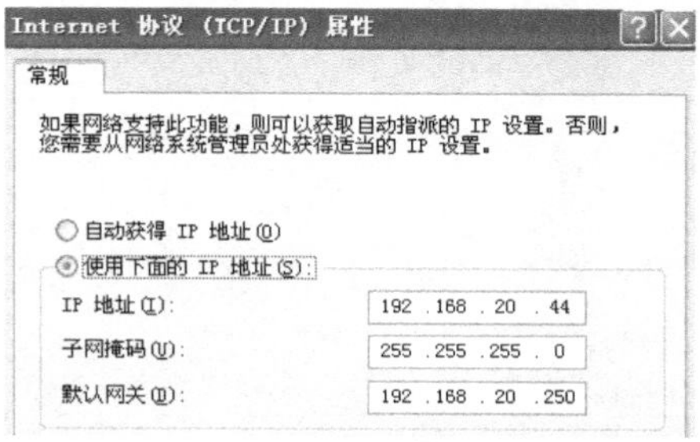

图 13-3 测试 PC 的相关设置（2）

多层交换技术（实践篇）

在系统中运行"C:\>tracert 172.16.6.1"命令，结果如下。

```
Tracing route to 172.16.6.1 over a maximum of 30 hops
  1    1 ms     <1 ms    <1 ms   192.168.20.254  // VLAN 20 的 IP 地址
  2    <1 ms    <1 ms    <1 ms   172.16.6.1
Trace complete.
```

可见，VLAN 20 的数据是通过 Switch B 传输的。

任务 14 配置交换端口二层聚合

【任务描述】

某公司的网络中心为了实现接入网络稳定性，在接入交换机和汇聚交换机的连接链路上使用了多条冗余链路，同时，为了增加带宽，多条冗余链路之间实施端口聚合，提升骨干链路的带宽，这样可以实现链路之间的冗余和备份效果，避免因骨干链路上的单点故障而导致网络中断。

【任务目标】

理解端口聚合的工作原理，掌握如何在交换机上配置端口聚合。

【组网拓扑】

图 14-1 所示网络拓扑为接入和汇聚交换机连接场景，两台骨干交换机之间通过冗余链路实现了冗余和备份。为增强骨干链路的稳定性，实现端口聚合，将聚合端口设置为 Trunk 模式，使用源 MAC 关键字实现流量平衡算法。

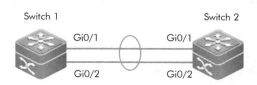

图 14-1 接入和汇聚交换机连接场景
（备注：具体的设备连接接口信息可根据实际情况决定）

【设备清单】

二层交换机（1 台），三层交换机（1 台），网线（若干），PC（若干）。

【关键技术】

链路聚合（Link Aggregation）是将多个物理端口汇聚在一起形成一个逻辑端口，以实现出/入流量吞吐量在各成员端口上的负载均衡。可以根据用户配置的端口实施负载均衡策略，决定封装完成的网络数据帧从哪个成员端口发送到对端的交换机。

当交换机检测到其中一个成员端口上的链路发生故障时，就停止在此端口上发送数据包，并根据负载均衡策略在剩下的链路中重新计算报文的发送端口，等故障端口恢复后，将再次担任收发端口。链路聚合技术在增加链路带宽、实现链路传输弹性和工程冗余等方面是一项很重要的技术。

交换机的端口聚合通常有两种模式，根据链路汇聚控制协议（Link Aggregation Control

多层交换技术（实践篇）

Protocol，LACP）不同，分为静态聚合（静态 LACP）模式和动态聚合（动态 LACP）模式，两模式的特点如下。

（1）在静态聚合模式下，聚合组内的各成员端口不启用任何协议，其端口状态（加入或离开）完全依据手工指定才能生效。

（2）在动态聚合模式下，聚合组内的各成员端口上均启用 LACP，其端口状态（加入或离开）通过该协议自动进行维护。

当交换端口启用 LACP 后，端口通过发送 LACPDU（Link Aggregation Control Protocol Data Unit，链路聚合控制协议数据单元）报文来通告自己的系统优先级、系统 MAC 地址、端口优先级、端口号和操作 KEY 等。相连的对端设备收到该报文后，选择端口进行相应的聚合操作，使双方在端口退出或者加入聚合组时保持一致。

实施端口聚合时，聚合端口通常呈现三种模式，分别是主动（Active）模式、被动（Passive）模式和静态模式。其中，主动模式的端口会主动发起 LACP 报文协商；被动模式的端口只会对收到的 LACP 报文做应答；静态模式不会发出 LACP 报文进行协商。

静态与 LACP 两种聚合方式的区别可以理解为静态路由与动态 OSPF 路由的区别，静态聚合方式根据管理员配置的方式强制生效；LACP 聚合方式通过协议报文与邻居协商状态，动态维护邻居关系与路由条目。

动态聚合可以动态发现链路故障，避免静态聚合时，因单条成员线路不通（如交换机端口虽然是 UP，但是由于中间光纤问题而不能通信）而导致异常，适用于用户要求可靠性较高，要求成员口动态加入离开的切换速度较快等情况。

但 LACP 在运行时会消耗设备资源，设备工作在 VSU 环境下且存在大量聚合口时，推荐使用静态聚合方式。

【实施步骤】

（1）在交换机 Switch 1 上配置聚合接口。

```
Ruijie#configure terminal
Ruijie(config)#hostname Switch 1
Switch 1(config)#interface range GigabitEthernet 0/1-2
                                   // 进入Gi0/1-2 接口配置模式
Switch 1(config-if-range)#port-group 1        // 创建静态聚合端口 AP1
       // 在接口模式下，使用 "port-group" 命令向聚合端口添加成员接口
       // 在接口配置模式下，使用 "no port-group" 命令使此成员接口退出 AP 聚合端口
Switch 1 (config-if-range)#exit
```

（2）在交换机 Switch 2 上配置 AP 聚合端口。

```
Ruijie#configure terminal
Ruijie(config)#hostname Switch 2
Switch 2 (config)#interface range GigabitEthernet 0/1-2
Switch 2 (config-if-range)#port-group 1        // 创建静态聚合端口 AP1
Switch 2 (config-if-range)#exit
```

（3）在交换机 Switch 1 上，将端口聚合设置为 Trunk 模式，设置流量平衡算法。

任务 ⑭　配置交换端口二层聚合

```
Switch 1 (config)#
Switch 1 (config)#interface AggregatePort 1           //进入AP1接口配置模式
Switch 1 (config-if-AggregatePort 1)#switchport mode trunk
Switch 1 (config-if-AggregatePort 1)#exit
Switch 1 (config)#aggregateport member linktrap
                                            // 打开成员口的LinkTrap功能
Switch 1 (config)#interface Aggregateport 1
Switch 1 (config-if-AggregatePort 1)#no snmp trap link-status
           // 关闭AP1聚合端口上的LinkTrap功能，实施安全的链路状态
Switch 1 (config-if-AggregatePort 1)#exit


Switch 1 (config)#aggregateport load-balance src-mac
    /* 更改AP1聚合端口上的流量平衡算法为源MAC模式，默认为源MAC+目的MAC模式。在不同
场景中，由于用户流量的特征不同，可能并不能使流量负载均衡到其成员链路上，此时需要人工调整负
载均衡方式 */
    /* 锐捷网络RGNOS的11.x版本支持对某个聚合端口指定负载均衡模式,对聚合端口指定了均衡
模式时，聚合端口的指定均衡模式将生效，不受全局均衡模式的影响，当聚合端口没有指定负载均衡模
式时，将采用全局负载均衡模式 */
```

（4）在交换机 Switch 2 上，将端口聚合设置为 Trunk 模式，设置流量平衡算法。

```
Switch 2 (config)#interface AggregatePort 1
Switch 2 (config-if-AggregatePort 1)#switchport mode trunk
Switch 2 (config-if-AggregatePort 1)#exit
Switch 2 (config)#aggregateport load-balance src-mac
                 // 更改流量平衡算法为源MAC模式，默认为源MAC+目的MAC模式
Switch 2 (config)#exit
```

（5）在交换机 Switch 1 上设置二层动态链路聚合（可选配置）。

```
Switch 1 (config)#
Switch 1 (config)#interface AggregatePort 1      // 进入AP1接口配置模式
Switch 1 (config-if-AggregatePort 1)#switchport mode trunk
Switch 1 (config-if-AggregatePort 1)#exit
Switch 1 (config)#interface range GigabitEthernet 0/1-2
Switch 1 (config-if-range)#port-group 1 mode active
                                     // 创建AP1接口,并设置模式为Active
Switch 1 (config-if-range)#exit
```

（6）在交换机 Switch 2 上设置二层动态链路聚合（可选配置）。

```
Switch 2 (config)#interface AggregatePort 1      // 进入AP1接口配置模式
Switch 2 (config-if-AggregatePort 1)#switchport mode trunk
Switch 2 (config-if-AggregatePort 1)#exit

Switch 2 (config)#interface range GigabitEthernet 0/1-2
Switch 2 (config-if-range)#port-group 1 mode active
```

多层交换技术（实践篇）

```
                                        // 创建 AP1 口，并设置模式为 Active
Switch 2 (config-if-range)#exit
```

（7）在交换机 Switch 1 上验证端口汇聚信息。

在交换机 Switch 1 上查看端口汇聚信息，聚合端口已经设置为 Trunk 模式，聚合组包括 Gi0/1 和 Gi0/2，如图 14-2 所示。

图 14-2　查看端口汇聚信息

在交换机 Switch 1 上，使用"show interface aggregatePort 1"命令，查看 AP1 端口信息，Gi0/1-2 单端口带宽是 1000Mbit/s，经过聚合后是 2000Mbit/s，表示端口已经聚合。

在交换机 Switch 1 上，使用"show aggregatePort load-balance""show interface aggregatePort 1"命令，查看聚合端口信息和流量平衡方式，可以看出流量平衡算法使用的是源 MAC 关键字，如图 14-3 所示。

```
Ruijie(config)#show aggregatePort load-balance
Load-balance     : Source MAC
Ruijie(config)#

    Vlan id: 1
Aggregate Port Informations:
    Aggregate Number: 1
    Name: "AggregatePort 1"
    Members: (count=2)
    GigabitEthernet 0/1                     Link Status: Up
    GigabitEthernet 0/2                     Link Status: Up
Load Balance by: Source MAC and Destination MAC
```

图 14-3　查看聚合端口信息和流量平衡方式

【备注】

在实施端口聚合时，需要注意以下几点。

（1）只有同类型端口才能聚合为一个 AP 端口，所有物理接口必须属于同一个 VLAN。

（2）端口聚合的成员属性必须一致，包括接口速率、双工、介质类型（指光口或者电口）等，光口和电口不能绑定，吉比特与万兆也不能绑定。

（3）二层端口只能加入二层 AP 聚合端口，三层端口只能加入三层 AP 聚合端口，已经关联了成员接口的 AP 聚合端口不允许改变二/三层属性。

（4）端口聚合后，成员口不单独配置，只能在 AP 聚合端口配置所需要功能（interface aggregateport x/x）。

（5）两个互连设备的端口聚合模式必须一致，并且同一时刻只能选择一种模式。

（6）设置流量平衡算法使用"源 IP+目的 IP"关键字。

```
Ruijie(config)#aggregateport load-balance src-dst-ip
```

任务 15 配置三层端口静态聚合

【任务描述】

某公司为了保障网络中心运行的稳定性,使用两台万兆交换机实现冗余和备份。两台交换机使用万兆光纤实现互连,不仅增加了链路带宽,还提高了网络可靠性。但为了实现三层网络的互连互通,还需要在两台核心设备之间运行三层静态链路聚合,实施基于"源IP+目的IP"关键字的流量平衡。

【任务目标】

配置三层端口静态聚合。

【组网拓扑】

图 15-1 所示网络拓扑为某企业核心网络场景。在两台三层汇聚交换机之间连接的冗余链路上实现端口聚合(Aggregate Port,AP),将聚合端口更改为三层端口,并配置 IP 地址,流量平衡算法使用"源 IP+目的 IP"关键字。

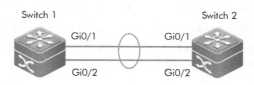

图 15-1 某企业核心网络场景

(备注:具体的设备连接接口信息可根据实际情况决定)

【设备清单】

三层交换机(两台),网线(若干),PC(若干)。

【关键技术】

将互连的多个物理端口聚合为一个逻辑端口,可以提高交换机之间的传输带宽,实现骨干链路冗余备份。为了充分利用现有聚合端口来实现高带宽传输,可以将聚合链路配置成 Trunk 端口。如果干道口中的一个端口发生堵塞,IP 数据包会被均衡地分配到该干道中的其他端口上传输,直到这个端口恢复正常,IP 数据包才被重新分配到该干道中所有的端口传输。

交换机不仅支持二层端口聚合,还支持三层端口聚合。二层端口和三层端口聚合的区别在于:二层端口不能配置 IP 地址,不能运行路由协议,只能对二层以太网帧进行转发;但三层端口可以配置 IP 地址,可运行路由协议,能接收并转发 IP 包。

一个聚合端口内的所有端口都必须具有相同的速率和双工模式。

多层交换技术（实践篇）

【实施步骤】

（1）在交换机 Switch 1 上创建聚合端口，将其更改为三层端口，配置 IP 地址。

① 创建 AP 聚合端口。

```
Ruijie#configure terminal
Ruijie(config)#interface range GigabitEthernet 0/1-2    // 进入 Gi0/1-2 接口
Ruijie(config-if-range)#port-group 1                    // 创建聚合端口 AP
Ruijie(config-if-range)#exit
```

② 将 AP 聚合端口设置为三层端口。

```
Ruijie(config)#interface aggregateport 1                // 进入创建的 AP 端口
Ruijie(config-if-AggregatePort 1)#no switchport         // 配置 AP 为三层端口
Ruijie(config-if-AggregatePort 1)#ip address 1.1.1.1 255.255.255.0
Ruijie(config-if-AggregatePort 1)#exit
```

（2）在交换机 Switch 2 上创建聚合端口，将其更改为三层端口，配置 IP 地址。

① 创建 AP 聚合端口。

```
Ruijie#configure terminal
Ruijie(config)# interface range GigabitEthernet 0/1-2   // 进入 Gi0/1-2 接口
Ruijie(config-if-range)#port-group 1                    // 创建聚合端口 AP
Ruijie(config-if-range)#exit
```

② 将 AP 聚合端口设置为三层端口。

```
Ruijie(config)# interface aggregateport 1               // 进入创建的 AP 端口
Ruijie(config-if-AggregatePort 1)#no switchport         // 配置 AP 为三层端口
Ruijie(config-if-AggregatePort 1)#ip address 1.1.1.2 255.255.255.0
Ruijie(config-if-AggregatePort 1)#exit
```

（3）在交换机 Switch 1 上设置三层聚合端口流量平衡算法。

```
Ruijie(config)#
Ruijie(config)#aggregateport load-balance src-dst-ip
                        // 设置流量平衡算法，使用"源 IP+目的 IP"关键字
Ruijie(config)#exit
```

（4）在交换机 Switch 2 上设置三层聚合端口流量平衡算法。

```
Ruijie(config)#
Ruijie(config)#aggregateport load-balance src-dst-ip
                        // 设置流量平衡算法，使用"源 IP+目的 IP"关键字
```

（5）在交换机 Switch 1 上验证端口汇聚信息。

在交换机 Switch 1 上查看端口汇聚信息，可见聚合端口已经设置为 Trunk 模式，聚合组包括 Gi0/1 和 Gi0/2，如图 15-2 所示。

图 15-2　查看端口汇聚信息

任务 ⑮ 配置三层端口静态聚合

在交换机 Switch 1 上使用"show interface aggregateport 1"命令查看 AP1 端口信息，可见 Gi0/1 接口原本是 1000Mbit/s，经过聚合后是 2000Mbit/s，表示端口已经聚合，且聚合后的端口为三层聚合端口，配置的 IP 地址是 1.1.1.1/24，如图 15-3 所示。

```
Ruijie#show interfaces aggregateport 1
Index(dec):27 (hex):1b
AggregatePort 1 is UP  , line protocol is UP
Hardware is Aggregate Link AggregatePort, address is 1414.4b5b.b5cf
Interface address is: 1.1.1.1/24
ARP type: ARPA, ARP Timeout: 3600 seconds
  MTU 1500 bytes, BW 200000 Kbit
  Encapsulation protocol is Ethernet-II, loopback not set
  Keepalive interval is 10 sec , set
  Carrier delay is 2 sec
  Rxload is 1/255, Txload is 1/255
```

图 15-3　聚合后的端口是三层聚合端口

在交换机 Switch 1 上查看流量平衡方式，可见使用基于"源 IP+目的 IP"关键字的算法实现了流量平衡，如图 15-4 所示。

```
Ruijie#show aggregatePort load-balance
Load-balance    : Source IP and Destination IP
Ruijie#
```

图 15-4　查看流量平衡方式

任务 16 配置三层端口动态聚合

【任务描述】

某公司的网络中心为了实现核心网络的稳定性和高带宽需求，在核心网络中实施了冗余和备份，通过端口聚合提升骨干链路的带宽，避免单点链路故障影响网络传输。此外，为了保障核心网络的健壮性，减少网络的人工运维工作量，实施了动态链路聚合技术。当聚合链路中的一条成员链路断开时，LACP 会将该成员链路的上流量自动分配到 AG 中的其他成员链路上。

【任务目标】

配置三层端口上的动态链路聚合。

【组网拓扑】

图 16-1 所示网络拓扑为某企业网络核心交换机连接场景，在核心交换机的骨干链路上实现了冗余和备份，并在骨干链路上实现了端口聚合，通过实施动态链路聚合技术实现聚合链路上的流量自动平衡算法。

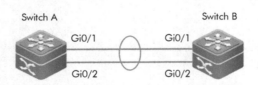

图 16-1 某企业网络核心交换机连接场景

（备注：具体的设备连接接口信息可根据实际情况决定）

【设备清单】

三层交换机（两台），网线（若干），PC（若干）。

【关键技术】

传统的以太网链路聚合是纯静态方式，两端的聚合设备之间不需要沟通和传播信息，数据帧在转发时，本地交换机只是依照自己的链路聚合算法，将流量分配在每条静态聚合链路上，对端交换机无论从哪条聚合链路上收到数据帧，都统一认为其来自于聚合端口。

纯静态的链路聚合方式的优点和缺点都非常明显：优点是不存在兼容性问题，端口聚合只是本地属性，与远端无关，对聚合端口的配置要求也较低，甚至有可能将不同速率的端口聚合在一起；缺点是由于两端设备互相不交换信息，如果在设备间的中间环节（如光纤收发

任务⑯ 配置三层端口动态聚合

器、协议转换器、安全设备等）出现问题，或者存在光纤单向通信的情况，会出现交换机无法感知链路故障，仍然向已经中断的线路分发流量的现象，这种部分聚合端口无法正常工作后流量仍然沿着链路转发的情况，会造成部分节点通信中断（具体要看交换机负载分担的算法），出现这种故障，对于网管人员来说是比较棘手的，因为故障比较隐蔽。

基于 IEEE 802.3ad 标准的链路聚合控制协议（Link Aggregation Control Protocol，LACP）能实现链路动态聚合，运行该协议的交换机之间通过互发链路聚合控制协议数据单元来交换链路聚合的相关信息，实现如下主要功能。

（1）使用 LACP 技术后，两端交换机可以交换链路信息，如果某条线路出现问题或者出现单向通信情况，信息交换就会停止，等待计时器超时后，该条问题线路就会被从聚合组中删除。

（2）使用 LACP 技术，只需要在端口上开启 LACP，不需要单独建立聚合组，端口连接后会根据收到的 LACPDU 信息实现自动聚合。例如，交换机 A 需要与交换机 B、C 分别建立两条线路的聚合连接时，只需要在交换机 A 上开启 4 个端口的 LACP，在 B 和 C 上分别开启两个端口的 LACP，进行物理连接即可，即不需要指定交换机 A 的哪两个接口对应交换机 B，哪两个接口对应交换机 C。

使用 LACP 聚合可以动态发现链路故障，避免静态链路聚合时单条成员线路不通（如交换机端口虽然是 UP 的，但是由于中间光纤出现问题而无法通信）导致的异常。

动态聚合协议会消耗设备资源，在核心层的设备进行互连时，或在 VSU 环境下存在大量聚合接口时，推荐使用静态聚合方式。

【实施步骤】

（1）在交换机 Switch A 上创建 AP 聚合端口。

① 创建 AP 聚合端口。

```
Ruijie#configure terminal
Ruijie(config)#hostname Switch A
Switch A(config)#interface range GigabitEthernet 0/1-2
Switch A(config-if-range)#port-group 3                    // 创建AP3聚合端口
Switch A(config-if-range)#exit
Switch A(config)#
```

② 配置创建的 AP 聚合端口为动态模式。

```
Switch A(config)#
Switch A(config)#interface range GigabitEthernet 0/1-2
Switch A(config-if-range)#port-group 3 mode active
            // 创建聚合端口 AP3，设置模式为 Active，表示端口以主动模式加入动态聚合组
               // 如果选择使用 Passive 模式，则表示端口以被动模式加入聚合组
Switch A(config-if-range)# exit
```

```
Switch A(config)# lacp system-priority 4096
    // 配置 LACP 的系统优先级，LACP 的系统优先级可选范围为 0～65535，默认优先级为 32768
（可选配置）
```

多层交换技术（实践篇）

　　/* 在全局模式下配置 LACP 的系统优先级。一台设备的所有动态链路组只能有一个 LACP 系统优先级，修改这个值会影响到交换机上的所有聚合组。在接口配置模式下，使用 "no lacp system-priority" 命令可将 LACP 的系统优先级恢复为默认值。对设备系统 ID 优先级进行配置时，配置值越小，系统 ID 优先级越高，系统 ID 优先级高的设备优先选择聚合端口 */

③ 在交换机 Switch A 上手工开启成员口的 LinkTrAG 通告功能。

```
Switch A(config)#
Switch A(config)#aggregateport member linktrAG
                                          // 成员接口 LinkTrAG 功能默认关闭
Switch A(config)#interface Aggregateport 3
Switch A(config-if-AggregatePort 3)#no snmp trAG link-status
```

　　/* 在接口模式下，用户可以对指定的 AP 口设置是否开启 LinkTrAG 通告功能。当该功能打开后，AP 口发生 Link 状态变化时将发出 LinkTrAG 通告，反之则不发出通告。默认情况下，该功能是打开的。用户可以在指定 AP 口的接口模式下，通过配置 "no snmp trAG link-status" 命令关闭指定 AP 口的 LinkTrAG 通告功能 */

（2）在交换机 Switch B 上创建 AP 聚合端口。

① 创建 AP 聚合端口。

```
Ruijie#configure terminal
Ruijie(config)#hostname Switch B
Switch B(config)#interface range GigabitEthernet 0/1-2
Switch B(config-if-range)#port-group 3                    // 创建聚合端口 AP3
Switch B(config-if-range)#exit
Switch B(config)#
```

② 配置创建的 AP 聚合端口为动态模式。

```
Switch B(config)#
Switch B(config)#interface range GigabitEthernet 0/1-2
Switch B(config-if-range)# port-group 3 mode active
        // 设置创建的聚合端口 AP3 的模式为 Active，表示端口以主动模式加入动态聚合组
Switch B(config-if-range)#exit
Switch B(config)# lacp system-priority 61440   // 配置 LACP 的系统优先级
```

（3）在交换机 Switch A 上查看动态聚合端口配置信息。

① 查看 AP 口和成员口的对应关系。

```
Switch A#show aggregateport summary
AggregatePort   MaxPorts   SwitchPort   Mode      Ports
-------------   --------   ----------   ------    -------------------------
Ag3             8          Enabled      ACCESS    Gi0/1, Gi0/2
```

② 查看 AP 口的流量均衡算法配置。

```
Switch A#show run | include AggregatePort 3
Building configuration...
Current configuration: 54 bytes
```

任务 ⓰ 配置三层端口动态聚合

```
interface AggregatePort 3
no snmp trAG link-status
```

```
Switch A#show run | include AggregatePort
aggregateport member linktrAG
```

```
Switch A#show aggregatePort load-balance      // 查看 AP 口的流量均衡算法配置
Load-balance : Source MAC
```

③ 查看 LACP 和成员口的对应关系是否正确。

```
Switch A#show LACP summary 3
System Id:32768, 00d0.f8fb.0001
Flags: S - Device is requesting Slow LACPDUs
F - Device is requesting Fast LACPDUs.
A - Device is in active mode. P - Device is in passive mode.
Aggregate port 3:
Local information:
                LACP port    Oper    Port   Port
Port   Flags    State Priority   Key   Number  State
-----------------------------------------------------------------
Gi0/1  SA  bndl  32768  0x3  0x1 0x3d
Gi0/2  SA  bndl  32768  0x3  0x2 0x3d
Partner information:
                LACP port    Oper    Port   Port
Port   Flags    Priority   Dev ID     Key   NumberState
-----------------------------------------------------------------
Gi0/1  SA  32768  00d0.f800.0002  0x3  0x1 0x3d
Gi0/2  SA  32768  00d0.f800.0002  0x3  0x2 0x3d
```

（4）在交换机 Switch B 上查看聚合端口配置信息。

① 查看 AP 口和成员口的对应关系。

```
Switch B#show aggregateport summary
AggregatePort    MaxPorts    SwitchPort    Mode       Ports
-------------    --------    ----------    ------     -----------------------------
Ag3                 8         Enabled     ACCESS    Gi0/1,Gi0/2
```

② 查看 AP 口的流量均衡算法配置。

```
Switch B#show run | include AggregatePort 3
Building configuration…
Current configuration: 54 bytes
interface AggregatePort 3
no snmp trAG link-status
```

多层交换技术（实践篇）

```
Switch B#show run | include AggregatePort
aggregateport member linktrAG
```

```
Switch B#show aggregatePort load-balance      // 查看AP口的流量均衡算法配置
Load-balance : Destination MAC
```

③ 查看LACP和成员口的对应关系是否正确。

```
Switch B#show LACP summary 3
System Id:32768, 00d0.f8fb.0002
Flags: S - Device is requesting Slow LACPDUs
F - Device is requesting Fast LACPDUs.
A - Device is in active mode. P - Device is in passive mode.
Aggregate port 3:
Local information:
             LACP port    Oper   Port   Port
Port  Flags  State  Priority  Key  Number  State
----------------------------------------------------------------
Gi0/1  SA   bndl   32768    0x3   0x1   0x3d
Gi0/2  SA   bndl   32768    0x3   0x2   0x3d
Partner information:
             LACP port    Oper   Port   Port
Port  Flags  Priority  Dev ID  Key  Number State
----------------------------------------------------------------
Gi1/1  SA  32768  00d0.f800.0001  0x3  0x1  0x3d
Gi1/2  SA  32768  00d0.f800.0001  0x3  0x2  0x3d
```

任务 ⑰ 配置计算机自动获取 IP 地址

【任务描述】

某公司为了减少手工配置内网中主机 IP 地址的工作量，决定使用 DHCP 来动态地为主机分配 IP 地址。由于公司的规模小，为了降低成本，不再搭建专门的 DHCP 服务器，利用现有路由器实现 DHCP 服务。其中，网段为 172.16.1.0/24，网关为 172.16.1.254，域名为 ruijie.com.cn，服务器农场（Server Farm，类似 DMZ 区域）中域名服务器 IP 地址规划为 172.16.1.253/24，WINS 服务器 IP 地址为 172.16.1.252/24，NETBIOS 节点类型为复合型，地址租期为 7 天，并要求给主机 MAC 地址为 0001.0001.0001 的设备分配 IP 地址 172.16.1.10/24。此外，该地址池中的 172.16.1.200/24～172.16.1.254/24 地址段不允许分配给客户端。

【任务目标】

配置 DHCP 单地址池，使出口路由器提供 DHCP 服务。

【组网拓扑】

图 17-1 所示网络拓扑为某公司网络内部的连接场景，使内网中的出口路由器承担 DHCP 服务，完成 DHCP 地址池、排除地址和静态地址绑定。

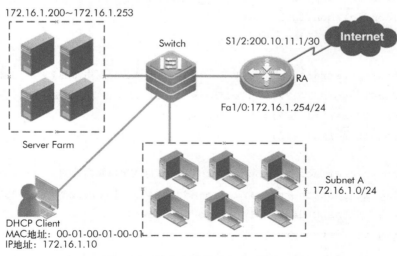

图 17-1 某公司网络内部的连接场景

（备注：具体的设备连接接口信息可根据实际情况决定）

【设备清单】

路由器（1 台），交换机（1 台），网线（若干），PC（若干）。

多层交换技术（实践篇）

【关键技术】

动态主机配置协议（Dynamic Host Configuration Protocol，DHCP）是局域网内的动态地址获取网络协议，使用 UDP 工作。DHCP 主要有两个用途：一是给内部网络中主机或网络服务器自动分配 IP 地址；二是为内网中需要动态配置的主机分配 IP 地址和提供主机配置参数，并设定地址绑定以及配置主机能够使用的地址的租期。

DHCP 采用客户端/服务器模型，主机地址的动态分配任务由网络中的 DHCP 服务器完成。当 DHCP 服务器接收到来自内部网络中主机申请 IP 地址的请求信息时，就会向请求 IP 地址的主机发送相关的地址配置等信息。

DHCP 有 3 种 IP 地址分配机制，分别如下。

（1）自动分配（Automatic Allocation）方式。在自动分配方式下，DHCP 服务器为网络中的主机指定一个永久性的 IP 地址，一旦 DHCP 客户端第一次成功从 DHCP 服务器端租用到 IP 地址，就可以永久地使用该地址。

（2）动态分配（Dynamic Allocation）方式。在动态分配方式下，DHCP 服务器给网络中的主机指定一个具有时间限制的 IP 地址，时间到期或主机明确表示放弃时，该地址可以被其他主机使用。

（3）手工分配（Manual Allocation）方式。在手工分配方式下，客户端的 IP 地址是由网络管理员手工指定的，DHCP 服务器只是将指定的 IP 地址分配给客户端主机。

在这三种地址分配方式中，只有动态分配方式可以重复使用客户端不再需要的 IP 地址。

【实施步骤】

（1）在路由器上配置 IP 地址和路由。

```
Ruijie#config terminal
Ruijie(config)#configure terminal
Ruijie(config)#hostname RA

RA(config)#interface serial 1/2          //进入出口路由器的 WAN 口，配置 IP 地址
RA(config-if-serial 1/2)#ip address 200.10.11.1 255.255.255.252
RA(config-if-serial 1/2)#no shutdown
RA(config-if-serial 1/2)#exit

RA(config)#interface FastEthernet 1/0    //进入出口路由器的 LAN 口，配置 IP 地址
RA(config-if-FastEthernet 1/0)#ip address 172.16.1.254 255.255.255.0
RA(config-if-FastEthernet 1/0)# no shutdown
RA(config-if-FastEthernet 1/0)#exit

RA(config)#ip route 0.0.0.0 0.0.0.0 serial 1/2      // 配置出口默认路由
```

（2）配置 DHCP 地址池。

```
RA(config)#service dhcp                              // 开启 DHCP 服务器
RA(config)#ip dhcp pool LAN-IP                       // 配置内网地址池名称
```

任务 ⑰ 配置计算机自动获取 IP 地址

```
RA(dhcp-config)#network 172.16.1.0 255.255.255.0    // 配置DHCP 服务器地址池
RA(dhcp-config)#default-router 172.16.1.254         // 配置默认网关地址
RA(dhcp-config)#lease 7 0 0                         // 配置地址租期为 7 天

RA(dhcp-config)#netbios-node-type h-node            // 配置 DHCP 节点类型为复合型
RA(dhcp-config)#netbios-name-server 172.16.1.252    // 配置 WINS 服务器的地址
RA(dhcp-config)#domain-name ruijie.com.cn           // 配置 DHCP 服务器的域名
RA(dhcp-config)#dns-server 172.16.1.253 200.1.1.10  // 配置 DNS 服务器的地址
RA(dhcp-config)#exit
```

(3) 配置手工地址绑定和排除地址。

```
RA(config)#ip dhcp pool mac-ip
RA(dhcp-config)#hardware-address 0001.0001.0001     // 配置绑定的 MAC 地址
RA(dhcp-config)#host 172.16.1.10 255.255.255.0      // 配置绑定的 IP 地址
RA(dhcp-config)#netbios-node-type h-node
RA(dhcp-config)#netbios-name-server 172.16.1.252
RA(dhcp-config)#domain-name ruijie.com.cn
RA(dhcp-config)#dns-server 172.16.1.253 200.1.1.10
RA(dhcp-config)#default-router 172.16.1.254
RA(dhcp-config)#exit
RA(config)#ip dhcp excluded-address 172.16.1.200 172.16.1.254
                                                    // 配置DHCP 服务器地址池中排除的 IP 地址
RA(config)#end
```

【测试验证】

(1) 利用内网中的测试计算机进行 IP 地址获取测试,将计算机连接到交换机上,本地连接中将地址配置选项设置为 "自动获得 IP 地址",如图 17-2 所示,以自动获得地址。

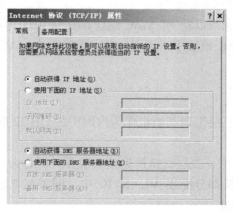

图 17-2　设置内网中的计算机自动获得 IP 地址

（2）在测试计算机的 DOS 命令行中，使用"ipconfig/all"命令查看主机 IP 地址，如图 17-3 所示，主机通过 DHCP 方式获取了指定网段的 IP 地址。

```
Connection-specific DNS Suffix  . : ruijie.com.cn
Description . . . . . . . . . . . : Intel(R) PRO/100 VE Network
Physical Address. . . . . . . . . : 00-16-D3-1F-2B-16
Dhcp Enabled. . . . . . . . . . . : Yes
Autoconfiguration Enabled . . . . : Yes
IP Address. . . . . . . . . . . . : 172.16.1.3
Subnet Mask . . . . . . . . . . . : 255.255.255.0
Default Gateway . . . . . . . . . : 172.16.1.254
DHCP Server . . . . . . . . . . . : 172.16.1.254
DNS Servers . . . . . . . . . . . : 172.16.1.253
                                    200.1.1.10
Primary WINS Server . . . . . . . : 172.16.1.252
```

图 17-3　查看主机 IP 地址（1）

（3）利用另一台测试计算机进行测试，把计算机连接到接入交换机上，本地连接中地址配置选项设置为"自动获得 IP 地址"。

（4）在 DOS 命令行中，使用"ipconfig /all"命令查看主机 IP 地址，如图 17-4 所示。

```
Connection-specific DNS Suffix  . : ruijie.com.cn
Description . . . . . . . . . . . : Intel(R) PRO/100 VE Network
Physical Address. . . . . . . . . : 00-16-D3-1F-2B-16
Dhcp Enabled. . . . . . . . . . . : Yes
Autoconfiguration Enabled . . . . : Yes
IP Address. . . . . . . . . . . . : 172.16.1.10
Subnet Mask . . . . . . . . . . . : 255.255.255.0
Default Gateway . . . . . . . . . : 172.16.1.254
DHCP Server . . . . . . . . . . . : 172.16.1.254
DNS Servers . . . . . . . . . . . : 172.16.1.253
                                    200.1.1.10
Primary WINS Server . . . . . . . : 172.16.1.252
```

图 17-4　查看主机 IP 地址（2）

（5）在路由器 RA 上，使用"show ip dhcp binding"命令验证手工地址绑定。

```
RA#show ip dhcp binding
IP address         Hardware address           Lease expiration           Type
172.16.1.10        0010.0010.0010             infinite                   Manual
```

从 show 命令的输出结果可以看到，MAC 地址 0010.0010.0010 绑定了 IP 地址 172.16.1.10。

备注：如果要测试 MAC 地址手工绑定，应注意配置的 MAC 地址为测试计算机的真实 MAC 地址。

【注意事项】

企业网中更多地使用三层汇聚交换机作为内网中的 DHCP 服务器，为本地网络中的终端设备提供 DHCP 服务。

在图 17-5 所示的场景中，二层交换机做接入服务，核心为三层汇聚交换机。其中，二层接入交换机下连的计算机都分配在 VLAN 10 中，网关在三层汇聚交换机上，全网使用 DHCP 方式动态获取 IP 地址。

任务 ⑰ 配置计算机自动获取 IP 地址

图 17-5 三层汇聚交换机作为 DHCP 服务器

其具体配置过程如下。

（1）在三层汇聚交换机上配置用户的网关地址。

```
Ruijie(config)#vlan 10
Ruijie(config)#interface vlan 10
Ruijie(config-if-VLAN 10)#ip address 192.168.1.254 255.255.255.0
Ruijie(config-if-VLAN 10)#exit
Ruijie(config)#interface GigabitEthernet 0/24
Ruijie(config-if-GigabitEthernet 0/24)#switch mode trunk
                            // 连接二层接入交换机接口封装为干道接口
```

（2）在三层汇聚交换机上配置 DHCP 地址池。

```
Ruijie(config)#
Ruijie(config)#service dhcp
            // 开启三层汇聚交换机的 DHCP 服务，该服务默认不启用
Ruijie(config)#ip dhcp pool vlan10-IP
Ruijie(dhcp-config)#network 192.168.1.0 255.255.255.0
                // 子网掩码要和所设置 IP 地址的子网掩码一致
Ruijie(dhcp-config)#dns-server 218.85.157.99   // 设置分配给客户端的 DNS 地址
Ruijie(dhcp-config)#default-router 192.168.1.254
// 分配给用户的网关地址
Ruijie(dhcp-config)#end
```

（3）在二层接入交换机上配置用户 VLAN 信息。

```
Ruijie(config)#
Ruijie(config)#interface range GigabitEthernet 0/1-22
Ruijie(config-if-range)#switch access vlan 10    //将用户端口划分到 VLAN 10 中
Ruijie(config-if-range)#exit
Ruijie(config)#interface GigabitEthernet 0/24
Ruijie(config-if-fastEthernet0/24)#switch mode trunk
                            // 接入交换机的接口封装为干道接口
Ruijie(config-if-fastEthernet0/24)#exit
```

多层交换技术（实践篇）

（4）在三层汇聚交换机上查看 DHCP 服务器地址池分配情况。

```
Ruijie#show ip dhcp binding
Total number of clients   : 2
Expired clients  : 0
Running clients  : 2
IP address      Hardware address      Lease expiration            Type
192.168.1.1     0021.cccf.6f70        000 days 23 hours 42 mins   Automatic
192.168.1.2     001a.a9c4.05f3        000 days 23 hours 42 mins   Automatic
/* IP address 字段（如"192.168.1.2"）为分配出去的 IP 地址；Lease expiration 字
段（如"000"）为剩余租约时间；Hardware address 字段（如"001a.a9c4.05f3"）为客户端 MAC
地址 */
```

（5）在测试计算机上查看获取 IP 地址的情况。在测试计算机 PC1 上进入 DOS 命令行，使用"ipconfig/all"命令，查看到图 17-6 所示的信息。

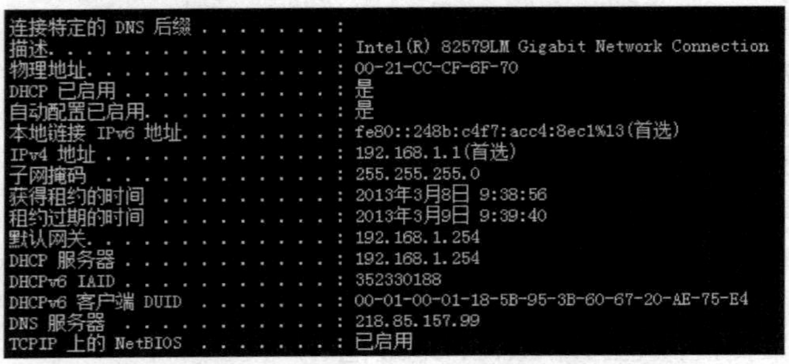

图 17-6　查看到的信息

任务 18 配置不同子网中设备 DHCP 自动获取地址

【任务描述】

某公司按照部门业务规划了 4 个部门网络，每个部门对应一个 VLAN，共创建 5 个 VLAN，分别分配一个子网。为了减少手工配置 IP 地址的麻烦，在三层汇聚交换机上配置 DHCP 服务，设置多个地址池群，各个部门 VLAN 中的主机均能通过 DHCP 自动申请到本 VLAN 网段相应的 IP 地址。

【任务目标】

配置 DHCP 的多地址池功能，实现基于不同 VLAN 的 DHCP 服务。

【组网拓扑】

图 18-1 所示网络拓扑为某公司的部门 VLAN 场景，通过三层汇聚交换机完成公司内部 DHCP 服务器配置。公司按部门共划分 5 个 VLAN 网段，其中 VLAN 10 为 192.168.10.0/24 网段；VLAN 20 为 192.168.20.0/24 网段；VLAN 30 为 192.168.30.0/24 网段；VLAN 40 为 192.168.40.0/24 网段；VLAN 50 为 192.168.50.0/24 网段。通过配置的 DHCP 服务使各个子网自动获取相应网段中的 IP 地址。

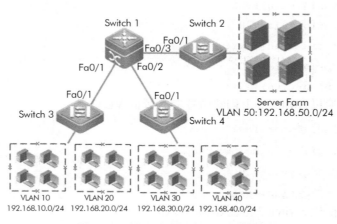

图 18-1 某公司的部门 VLAN 场景

（备注：具体的设备连接接口信息可根据具体情况决定）

其中，VLAN 50 是公司服务器区域群，需要静态指定 IP 地址。服务区域中 DNS 服务器的 IP 地址为 192.168.50.10/24，为公网提供的 DNS 服务器的 IP 地址为 200.1.1.10/24，域名为 ruijie.com.cn。

多层交换技术（实践篇）

此外，每个部门 VLAN 中尾数为 180～253 的地址不允许分配给客户端。

【设备清单】

二层交换机（3 台），三层交换机（1 台），网线（若干），PC（若干）。

【关键技术】

在三层交换机上创建多个 DHCP 地址池，定义 IP 地址池的网段与 SVI 接口的 IP 对应关系，即多个不同 IP 子网对应各个 SVI 接口，最终使不同部门 VLAN 网段中的客户端计算机自动获取相应 IP 子网中的地址。

各个 VLAN 中的主机向三层交换机上的 DHCP 服务器申请 IP 地址时，DHCP 服务器根据申请设备所在 VLAN 分配相应的子网 IP 地址，从而实现多子网地址池的 DHCP 配置。

【实施步骤】

（1）在交换机上创建 VLAN。

① 在三层交换机 Switch 1 上创建部门 VLAN。

```
Ruijie#configure terminal
Ruijie(config)#hostname Switch 1
Switch 1(config)#vlan 10       // 创建各个部门和服务器区域群 VLAN
Switch 1(config-vlan)#exit
Switch 1(config)#vlan 20
Switch 1(config-vlan)#exit
Switch 1(config)#vlan 30
Switch 1(config-vlan)#exit
Switch 1(config)#vlan 40
Switch 1(config-vlan)#exit
Switch 1(config)#vlan 50
Switch 1(config-vlan)#exit
```

② 在接入交换机 Switch 2 上创建服务器区域群 VLAN。

```
Ruijie#configure terminal
Ruijie(config)#hostname Switch 2
Switch 2(config)#vlan 50
Switch 2(config-vlan)#exit
```

③ 在接入交换机 Switch 3 上创建部门 VLAN。

```
Ruijie#configure terminal
Ruijie(config)#hostname Switch 3
Switch 3(config)#vlan 10
Switch 3(config-vlan)#exit
Switch 3(config)#vlan 20
```

任务⑱ 配置不同子网中设备 DHCP 自动获取地址

```
Switch 3(config-vlan)#exit
Switch 3(config)#vlan 30
Switch 3(config-vlan)#exit
Switch 3(config)#vlan 40
Switch 3(config-vlan)#exit
```

④ 在接入交换机 Switch 4 上创建部门 VLAN。

```
Ruijie#configure terminal
Ruijie(config)#hostname Switch 4
Switch 4(config)#vlan 10
Switch 4(config-vlan)#exit
Switch 4(config)#vlan 20
Switch 4(config-vlan)#exit
Switch 4(config)#vlan 30
Switch 4(config-vlan)#exit
Switch 4(config)#vlan 40
Switch 4(config-vlan)#exit
```

（2）在三层交换机 Switch 1 上配置 VLAN 的 SVI 网关 IP 地址。

```
Switch 1(Config)#
Switch 1(config)#interface vlan 10          // 配置 VLAN 10 的网关
Switch 1(config-if-vlan 10)#ip address 192.168.10.254 255.255.255.0
Switch 1(config-if-vlan 10)#exit
Switch 1(config)#interface vlan 20          // 配置 VLAN 20 的网关
Switch 1(config-if-vlan 20)#ip address 192.168.20.254 255.255.255.0
Switch 1(config-if-vlan 20)#exit
Switch 1(config)#interface vlan 30          // 配置 VLAN 30 的网关
Switch 1(config-if-vlan 30)#ip address 192.168.30.254 255.255.255.0
Switch 1(config-if-vlan 30)#exit
Switch 1(config)#interface vlan 40          // 配置 VLAN 40 的网关
Switch 1(config-if-vlan 40)#ip address 192.168.40.254 255.255.255.0
Switch 1(config-if-vlan 40)#exit
Switch 1(config)#interface vlan 50          // 配置 VLAN 50 的网关
Switch 1(config-if-vlan 50)#ip address 192.168.50.254 255.255.255.0
Switch 1(config-if-vlan 50)#exit
```

（3）在所有交换机上分配接口到划分的 VLAN 中，并配置 Trunk。

① 在三层交换机 Switch 1 上完成接口和干道配置。

```
Switch 1(config)#
Switch 1(config)#interface range FastEthernet 0/1-3
Switch 1(config-if-range)#switchport mode trunk
```

多层交换技术（实践篇）

```
Switch 1(config-if-range)#exit
Switch 1(config)#
```

② 在接入交换机 Switch 2 上完成 VLAN 及接口配置。

```
Switch 2(config)#
Switch 2(config)#interface range FastEthernet 0/1
Switch 2(config-if-FastEthernet 0/1)#switchport mode trunk
Switch 2(config-if-FastEthernet 0/1)#exit
Switch 2(config)#interface range FastEthernet 0/2-10   // 把接口划分到 VLAN 50 中
Switch 2(config-if-range)#switchport access vlan 50
Switch 2(config-if-range)#exit
```

③ 在接入交换机 Switch 3 上完成 VLAN 及接口配置。

```
Switch 3(config)#
Switch 3(config)#interface range FastEthernet 0/1
Switch 3(config-if-FastEthernet 0/1)#switchport mode trunk
Switch 3(config-if-FastEthernet 0/1)#exit
Switch 3(config)#interface range FastEthernet 0/2-10   // 把接口划分到 VLAN 10 中
Switch 3(config-if-range)#switchport access vlan 10
Switch 3(config-if-range)#exit
Switch 3(config)#interface range FastEthernet 0/11-24  // 把接口划分到 VLAN 20 中
Switch 3(config-if-range)#switchport access vlan 20
Switch 3(config-if-range)#exit
```

④ 在接入交换机 Switch 4 上完成 VLAN 及接口配置。

```
Switch 4(config)#
Switch 4(config)#interface range FastEthernet 0/1
Switch 4(config-if-FastEthernet 0/1)#switchport mode trunk
Switch 4(config-if-FastEthernet 0/1)#exit
Switch 4(config)#interface range FastEthernet 0/2-10   // 把接口划分到 VLAN 30 中
Switch 4(config-if-range)#switchport access vlan 30
Switch 4(config-if-range)#exit
Switch 4(config)#interface range FastEthernet 0/11-24  // 把接口划分到 VLAN 40 中
Switch 4(config-if-range)#switchport access vlan 40
Switch 4(config-if-range)#exit
```

（4）在三层交换机 Switch 1 上配置 DHCP 地址池和排除地址。

```
Switch 1(config)#server dhcp                              // 开启 DHCP 服务器
Switch 1(config)#ip dhcp pool VLAN10-ip                   // 配置 VLAN 10 的地址池
Switch 1(dhcp-config)#domain-name ruijie.com.cn
Switch 1(dhcp-config)#network 192.168.10.0 255.255.255.0
Switch 1(dhcp-config)#dns-server 192.168.50.10 200.1.1.10
Switch 1(dhcp-config)#default-router 192.168.10.254
Switch 1(dhcp-config)#exit
```

任务 ⑱ 配置不同子网中设备 DHCP 自动获取地址

```
Switch 1(config)#ip dhcp pool VLAN20-ip          // 配置 VLAN 20 的地址池
Switch 1(dhcp-config)#domain-name ruijie.com.cn
Switch 1(dhcp-config)#network 192.168.20.0 255.255.255.0
Switch 1(dhcp-config)#dns-server 192.168.50.10 200.1.1.10
Switch 1(dhcp-config)#default-router 192.168.20.254
Switch 1(dhcp-config)#exit

Switch 1(config)#ip dhcp pool VLAN30             // 配置 VLAN 30 的地址池
Switch 1(dhcp-config)#domain-name ruijie.com.cn
Switch 1(dhcp-config)#network 192.168.30.0 255.255.255.0
Switch 1(dhcp-config)#dns-server 192.168.50.10 200.1.1.10
Switch 1(dhcp-config)#default-router 192.168.30.254
Switch 1(dhcp-config)#exit

Switch 1(config)#ip dhcp pool VLAN40             // 配置 VLAN 40 的地址池
Switch 1(dhcp-config)#domain-name ruijie.com.cn
Switch 1(dhcp-config)#network 192.168.40.0 255.255.255.0
Switch 1(dhcp-config)#dns-server 192.168.50.10 200.1.1.10
Switch 1(dhcp-config)#default-router 192.168.40.254
Switch 1(dhcp-config)#exit

Switch 1(config)#ip dhcp excluded-address 192.168.10.180 192.168.10.253
Switch 1(config)#ip dhcp excluded-address 192.168.20.180 192.168.20.253
Switch 1(config)#ip dhcp excluded-address 192.168.30.180 192.168.30.253
Switch 1(config)#ip dhcp excluded-address 192.168.40.180 192.168.40.253
                                                 // 配置各个部门 VLAN 中的排除地址
```

【测试验证】

（1）将测试计算机连接到 VLAN 10 中的交换机端口上，将本地连接中的地址配置选项设置为"自动获得 IP 地址"。在 DOS 命令行中，使用"ipconfig/all"命令查看获取的 IP 地址信息，如图 18-2 所示，VLAN 10 网段主机自动获取到 VLAN 10 网段的 IP 地址。

```
Connection-specific DNS Suffix  . : ruijie.com.cn
Description . . . . . . . . . . . : Intel(R) PRO/100 VE Network
Physical Address. . . . . . . . . : 00-16-D3-1F-2B-16
Dhcp Enabled. . . . . . . . . . . : Yes
Autoconfiguration Enabled . . . . : Yes
IP Address. . . . . . . . . . . . : 192.168.10.1
Subnet Mask . . . . . . . . . . . : 255.255.255.0
Default Gateway . . . . . . . . . : 192.168.10.254
DHCP Server . . . . . . . . . . . : 192.168.10.254
DNS Servers . . . . . . . . . . . : 192.168.50.10
```

图 18-2 VLAN 10 网段主机自动获取到 IP 地址

多层交换技术（实践篇）

（2）将测试计算机连接到 VLAN 20 中的交换机端口上，将本地连接中地址配置选项设置为"自动获得 IP 地址"。在 DOS 命令行中，使用"ipconfig/all"命令查看获取的 IP 地址信息，如图 18-3 所示，VLAN 20 网段主机自动获取到 VLAN 20 网段的 IP 地址。

```
Connection-specific DNS Suffix  . : ruijie.com.cn
Description . . . . . . . . . . . : Intel(R) PRO/100 VE Network
Physical Address. . . . . . . . . : 00-16-D3-1F-2B-16
Dhcp Enabled. . . . . . . . . . . : Yes
Autoconfiguration Enabled . . . . : Yes
IP Address. . . . . . . . . . . . : 192.168.20.1
Subnet Mask . . . . . . . . . . . : 255.255.255.0
Default Gateway . . . . . . . . . : 192.168.20.254
DHCP Server . . . . . . . . . . . : 192.168.20.254
DNS Servers . . . . . . . . . . . : 192.168.50.10
                                    200.1.1.10
```

图 18-3　VLAN 20 网段主机自动获取到 IP 地址

任务 ⑲ 配置交换机 DHCP Relay

【任务描述】

某公司办公网在三层交换机上配置了 DHCP 技术，帮助办公网络中的所有计算机动态获取 IP 地址，实现每台设备终端自动获取指定网络的 IP 地址的功能。由于公司接入网络中终端设备众多，所以决定在网络中心部署一台专门的 DHCP Server（通常是 Windows 的服务器或者 Linux 的服务器），实现公司内网中 IP 地址的统一分配和维护。公司中其他部门都连接在接入交换机上，通过配置 DHCP Relay 功能，实现办公网络中 Client 与 Server 之间 DHCP 服务代理的交互。

【任务目标】

掌握配置 DHCP Relay 的方法。

【组网拓扑】

图 19-1 所示网络拓扑为某公司办公网络场景，在三层汇聚交换机 Switch 1 上开启 DHCP Server，二层接入交换机 Switch 2 上开启 DHCP Relay 功能，实现所连接用户终端设备自动获取 IP 地址的功能。

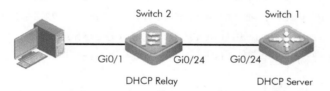

图 19-1　某公司办公网络场景
（备注：具体的设备连接接口信息可根据实际情况决定）

【设备清单】

二层交换机（1 台），三层交换机（1 台），网线（若干），PC（若干）。

【关键技术】

DHCP 广泛用于动态分配可重用的网络 IP 地址资源，DHCP 服务器为内网中需要动态配置的主机分配 IP 地址和提供主机配置参数（如网关、DNS Server 等）。首先，内网中的客户端发出 DISCOVER 广播报文给 DHCP 服务器；DHCP 服务器收到 DISCOVER 报文后，根据一定的策略来给客户端分配 IP 地址资源，发送出 OFFER 报文。其次，DHCP 客户端收到 OFFER 报文后，验证资源是否可用，如果资源可用，则发送 REQUEST 报文，如果资

多层交换技术（实践篇）

源不可用，则重新发送 DISCOVER 报文。再次，服务器收到 REQUEST 报文后，验证 IP 地址资源（或其他有限资源）是否可以分配，如果可以分配，则发送 ACK 报文，如果不可分配，则发送 NAK 报文。最后，如果 DHCP 客户端收到 ACK 报文，则开始使用服务器分配的资源；如果收到 NAK 报文，则可能重新发送 DISCOVER 报文。

DHCP Relay 的功能是在 DHCP 服务器和客户端之间转发 DHCP 数据包。当 DHCP 客户端与服务器不在同一个子网时，就必须由 DHCP Relay 来转发 DHCP 请求和应答消息。DHCP 中继代理中的数据转发与通常的路由转发是不同的。通常的路由转发相对来说是透明传输的，设备一般不会修改 IP 包内容；而 DHCP Relay 接收到 DHCP 消息后，会转换源/目的 IP，使用 MAC 地址生成一个 DHCP 消息，并转发出去。

在 DHCP 客户端看来，DHCP Relay 就像是一台 DHCP 服务器；在 DHCP 服务器看来，DHCP Relay 就像是一个 DHCP 客户端。DHCP Relay 将收到的 DHCP 广播请求报文以单播方式转发给 DHCP 服务器，并将收到的 DHCP 响应报文转发给 DHCP 客户端。

DHCP Relay 相当于一个转发站，负责沟通位于不同网段中的 DHCP 客户端和 DHCP 服务器。不需要在每个子网中都启用 DHCP Server，只要在网络中心安装一台 DHCP 服务器，就可以对多个不同网段实施动态 IP 管理，即实施 Client—Relay—Server 模式的 DHCP 动态 IP 管理。

【实施步骤】

（1）在三层汇聚交换机 Switch 1 上配置接口信息及 IP 地址。

```
Ruijie#configure terminal
Ruijie(config)#hostname Switch 1
Switch 1(config)#interface GigabitEthernet 0/24
Switch 1(config-if-GigabitEthernet 0/24)#no switchport
Switch 1(config-if-GigabitEthernet 0/24)#ip address 172.16.1.1 255.255.255.252
Switch 1(config-if-GigabitEthernet 0/24)#no shutdown
Switch 1(config-if-GigabitEthernet 0/24)#exit

Switch 1 (config)#ip route 192.168.1.0 255.255.255.0 172.16.1.2
                                              // 配置指向用户网络的静态路由
```

（2）在三层汇聚交换机 Switch 1 上开启 DHCP 服务器，配置 DHCP 地址池。

```
Switch 1(config)#service dhcp
Switch 1(config)#ip dhcp pool vlan10-IP
Switch 1(dhcp-config)#network 192.168.1.0 255.255.255.0
Switch 1(dhcp-config)#dns-server 8.8.8.8
Switch 1(dhcp-config)#default-router 192.168.1.254
Switch 1(dhcp-config)#exit
```

（3）在二层接入交换机 Switch 2 上配置用户 SVI 网关。

```
Ruijie#configure terminal
Ruijie(config)#hostname Switch 2
```

任务 ⑲ 配置交换机 DHCP Relay

```
Switch 2(config)#vlan 10
Switch 2(config-vlan)#exit
Switch 2(config)#interface VLAN 10
Switch 2(config-if-VLAN 10)#ip address 192.168.1.254 255.255.255.0
Switch 2(config-if-VLAN 10)#no shutdown
Switch 2(config-if-VLAN 10)#exit
```

（4）在二层接入交换机 Switch 2 上实现与 Switch 1 的互连。

```
Switch 2(config)#
Switch 2(config)#interface GigabitEthernet 0/24
Switch 2(config-if-GigabitEthernet 0/24)#no switchport
Switch 2(config-if-GigabitEthernet 0/24)#ip address 172.16.1.2 255.255.255.252
Switch 2(config-if-GigabitEthernet 0/24)#no shutdown
Switch 2(config-if-GigabitEthernet 0/24)#exit

Switch 2(config)#ip route 0.0.0.0 0.0.0.0 172.16.1.1
                                        // 配置指向 DHCP 服务器的默认路由
```

（5）在二层接入交换机 Switch 2 上配置 DHCP Relay。

```
Switch 2(config)#service dhcp            // 开启 DHCP 服务
Switch 2(config)#ip helper-address 172.16.1.1     // 开启 DHCP Relay 功能
```

【测试验证】

（1）查看三层汇聚交换机 Switch 1 上的 DHCP 服务器地址池分配。

```
Switch 1#show ip dhcp binding
Total number of clients    : 2
Expired clients            : 0
Running clients            : 2
IP address        Hardware address      Lease expiration           Type
192.168.1.1       0021.cccf.6f70        000 days 23 hours 42 mins  Automatic
192.168.1.2       001a.a9c4.05f3        000 days 23 hours 42 mins  Automatic
/* IP address 字段（如 "192.168.1.2"）为分配出去的 IP 地址；Lease expiration 字段（如 "000"）为剩余租约时间；Hardware address 字段（如 "001a.a9c4.05f3"）为客户端 MAC 地址 */
```

（2）在终端设备上查看地址信息。

① 根据图 19-1 所示的拓扑结构将设备连接好，将测试计算机的本地连接 IP 地址设置为自动获得。

② 在测试计算机的 DOS 命令行中，使用 "ipconfig/all" 命令查看自动获取的地址信息，如图 19-2 所示。

③ 在接入交换机 Switch 2 上使用如下命令，查看 Switch 2 的 DHCP Relay 转发报文的

多层交换技术（实践篇）

情况。

```
Switch2#show ip dhcp relay-statistics
```

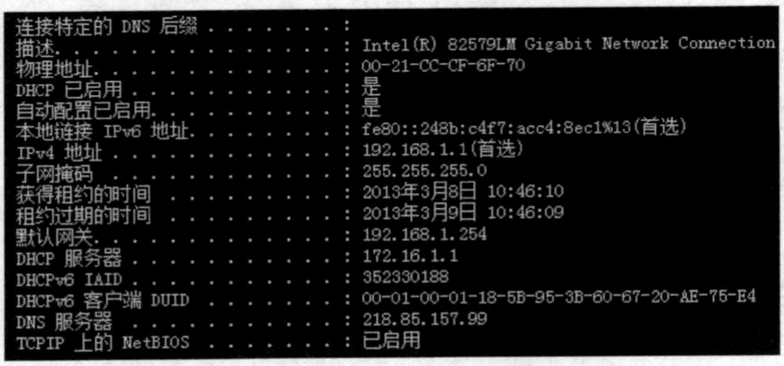

```
连接特定的 DNS 后缀 . . . . . . . :
描述. . . . . . . . . . . . . . . : Intel(R) 82579LM Gigabit Network Connection
物理地址. . . . . . . . . . . . . : 00-21-CC-CF-6F-70
DHCP 已启用 . . . . . . . . . . . : 是
自动配置已启用. . . . . . . . . . : 是
本地链接 IPv6 地址. . . . . . . . : fe80::248b:c4f7:acc4:8ec1%13(首选)
IPv4 地址 . . . . . . . . . . . . : 192.168.1.1(首选)
子网掩码  . . . . . . . . . . . . : 255.255.255.0
获得租约的时间 . . . . . . . . . . : 2013年3月8日 10:46:10
租约过期的时间 . . . . . . . . . . : 2013年3月9日 10:46:09
默认网关. . . . . . . . . . . . . : 192.168.1.254
DHCP 服务器 . . . . . . . . . . . : 172.16.1.1
DHCPv6 IAID . . . . . . . . . . . : 352330188
DHCPv6 客户端 DUID. . . . . . . . : 00-01-00-01-18-5B-95-3B-60-67-20-AE-75-E4
DNS 服务器 . . . . . . . . . . . . : 218.85.157.99
TCPIP 上的 NetBIOS . . . . . . . . : 已启用
```

图 19-2 查看自动获取的地址信息

任务 ⑳ 配置 DHCP Snooping 保障 DHCP 服务器的安全性

【任务描述】

某公司为减少网络管理工作量,在办公网络中搭建了 DHCP 服务器,以帮助终端设备自动获取 IP 地址。在日常使用中,有工作人员为个人移动上网方便,将自带家庭无线路由器连接在了办公网接口上,以扩展办公无线网络。家庭无线路由器上的 DHCP 地址获取功能,经常引导办公网络中的移动设备获取错误 IP 地址,导致用户获取到错误地址而不能上网或引发地址冲突。

在办公网络的 DHCP 服务器上,需要实施 DHCP Snooping(DHCP 监听)和 IP Source Guard 功能,以防止用户私设静态 IP 地址,并防止用户伪造 IP 地址进行内网扫描攻击。

【任务目标】

实施 DHCP Snooping 功能配置。

【组网拓扑】

图 20-1 所示拓扑为某公司网络办公场景,在接入交换机上开启 DHCP Snooping 功能,实施 DHCP Snooping,保证用户获得正确的 IP 地址。

图 20-1 某公司网络办公场景

(备注:具体的设备连接接口信息可根据实际情况决定)

【设备清单】

二层交换机(1 台),三层交换机(1 台),网线(若干),PC(若干)。

【关键技术】

DHCP Snooping 意为 DHCP 监听,在计算机动态获取 IP 地址的过程中,通过对客户端

多层交换技术（实践篇）

和服务器之间的 DHCP 交互报文进行监听，实现对用户的监控。同时，DHCP Snooping 起到了 DHCP 报文过滤的功能，通过合理的配置，实现对内网中非法服务器的过滤，防止客户端因获取到非法 DHCP 服务器提供的地址而无法上网。

当网络中存在 DHCP 服务器欺骗的时候，可以考虑采用这个功能，如网络中有个别终端使用的是 Windows 2003，或者 Windows 2008 默认开启了 DHCP 分配 IP 地址的服务，或者一些接入层的端口连接的 TP-Link、D-Link 等无线路由器开启了 DHCP 分配 IP 地址的服务。

推荐在用户接入层交换机上部署该功能，其越靠近计算机端口，控制越准确，建议每台交换机的接口只连接一台计算机，否则，如果交换机某个接口下串接一台 Hub，在 Hub 上连接若干台计算机，而凑巧该 Hub 下发生了 DHCP 欺骗，由于欺骗报文都在 Hub 接口之间直接转发，没有受到接入层交换机的 DHCP Snooping 功能控制，因此，这样的 DHCP 欺骗就无法防止。

【实施步骤】

（1）在三层汇聚交换机 Switch 1 上配置管理 IP 地址，即用户网关地址。

```
Ruijie(config)#
Ruijie(config)#interface VLAN 1
Ruijie(config-if-VLAN 1)#ip address 192.168.1.254 255.255.255.0
Ruijie(config-if-VLAN 1)#no shutdown
Ruijie(config-if-VLAN 1)#exit
```

（2）在三层汇聚交换机 Switch 1 上开启 DHCP 服务，创建 DHCP 地址池。

```
Ruijie(config)#service dhcp
Ruijie(config)#ip dhcp pool vlan1-IP
Ruijie(dhcp-config)#network 192.168.1.0 255.255.255.0    // 创建 DHCP 地址池
Ruijie(dhcp-config)#dns-server 218.85.157.99    // 设置分配给客户端的 DNS 地址
Ruijie(dhcp-config)#default-router 192.168.1.254
Ruijie(dhcp-config)#end
        // 分配用户的网关地址，和设备的 IP 地址一致，为 192.168.1.254
```

（3）在二层接入交换机 Switch 2 上开启 DHCP Snooping 功能。

```
Ruijie#configure terminal
Ruijie(config)#ip dhcp snooping        // 开启 DHCP Snooping 功能
```

（4）在二层接入交换机 Switch 2 上，配置连接 DHCP 服务器的接口为可信任接口。

```
Ruijie(config)#
Ruijie(config)#interface GigabitEthernet 0/24
Ruijie(config-GigabitEthernet 0/49)#ip dhcp snooping trust
/* 开启 DHCP Snooping 功能的交换机的所有接口默认为 untrust 接口，交换机只转发从 trust 接口收到的 DHCP 响应报文（如 OFFER、ACK 报文） */
```

【测试验证】

（1）在三层汇聚交换机 Switch 1 上查看 DHCP 服务器地址池分配情况，如图 20-2 所示。

```
Ruijie#show ip dhcp binding
IP address         Client-Identifier/      Lease expiration          Type
                   Hardware address        租约时间
192.168.1.1        0100.21cc.cf6f.70       000 days 23 hours 42 mins   Automatic
192.168.1.2        0100.1aa9.c405.f347.    000 days 23 hours 44 mins   Automatic
分配出IP地址         6967.6162.6974.4574.
                   6865.726e.6574.302f.    客户端标识，最前面2位数字表示以太网获取
                   31                      后面12位为客户端MAC地址
```

图 20-2　查看 DHCP 服务器地址池分配情况

（2）在测试计算机上，进入 DOS 命令行，使用“ipconfig/all”命令查看测试计算机获取的 IP 地址，如图 20-3 所示。

```
连接特定的 DNS 后缀 . . . . . . . :
描述. . . . . . . . . . . . . . . : Intel(R) 82579LM Gigabit Network Connection
物理地址. . . . . . . . . . . . . : 00-21-CC-CF-6F-70
DHCP 已启用 . . . . . . . . . . . : 是
自动配置已启用. . . . . . . . . . : 是
本地链接 IPv6 地址. . . . . . . . : fe80::248b:c4f7:acc4:8ec1%13(首选)
IPv4 地址 . . . . . . . . . . . . : 192.168.1.1(首选)
子网掩码 . . . . . . . . . . . . : 255.255.255.0
获得租约的时间 . . . . . . . . . : 2013年3月8日 9:38:56
租约过期的时间 . . . . . . . . . : 2013年3月9日 9:39:40
默认网关. . . . . . . . . . . . . : 192.168.1.254
DHCP 服务器 . . . . . . . . . . . : 192.168.1.254
DHCPv6 IAID . . . . . . . . . . . : 352330188
DHCPv6 客户端 DUID . . . . . . . : 00-01-00-01-18-5B-95-3B-60-67-20-AE-75-E4
DNS 服务器 . . . . . . . . . . . : 218.85.157.99
TCPIP 上的 NetBIOS . . . . . . . : 已启用
```

图 20-3　查看测试计算机获取的 IP 地址

（3）在三层汇聚交换机 Switch 1 上查看 DHCP Snooping 表，如图 20-4 所示。

```
Ruijie#show ip dhcp snooping binding

Total number of bindings: 2

MacAddress          IpAddress         Lease(sec)    Type            VLAN   Interface
------------------  ----------------  ------------  --------------  -----  ------------------
0021.cccf.6f70      192.168.1.1       86234         dhcp-snooping   1      FastEthernet 0/1
001a.a9c4.05f3      192.168.1.2       86367         dhcp-snooping   1      FastEthernet 0/2
```

图 20-4　查看 DHCP Snooping 表

（4）在二层接入交换机 Switch 2 上查看 DHCP Snooping 的设置情况，如图 20-5 所示。

```
Ruijie#show ip dhcp snooping

Switch DHCP snooping  status                        :   ENABLE
DHCP snooping  Verification of hwaddr status        :   DISABLE
DHCP snooping database write-delay time             :   0 seconds
DHCP snooping option 82 status                      :   DISABLE
DHCP snooping Support bootp bind status             :   DISABLE
Interface                         Trusted         Rate limit (pps)
------------------------------    -------         ----------------
GigabitEthernet 0/24              YES  信任端口    unlimited
```

图 20-5　查看 DHCP Snooping 的设置情况

任务 ㉑ 使用 RLDP 技术快速检测以太网链路故障

【任务描述】

某商贸大楼中，各个楼层的商家网络通过预留的接口接入大楼的网络。经常有商家在自己的店面随意扩展网络，不专业的安装导致大楼的网络出现环路。

由于店面众多，所以每次网络故障发生时，网络管理人员要对一家家店面进行排查，非常麻烦。为了避免以上情况发生，网络管理人员决定在接入端口上开启快速链路检测协议（Rapid Link Detection Protocol，RLDP）功能，以便快速检测以太网链路故障。RLDP 可以在接入交换机上实现环路检测，作为一种优化配置事先部署技术，可以有效防止接口下的各类环路问题，特别适合交换机下连集线器自身打环的情况（BPDU Guard 无法防止这类环路）。

【任务目标】

配置 RLDP 技术，快速检测以太网链路故障。

【组网拓扑】

图 21-1 所示网络拓扑为某商贸大楼部分店面网络接入场景。为了防止接入交换机下连的端口出现环路，需要在交换机上配置 RLDP 功能。

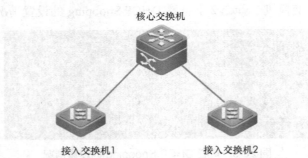

图 21-1 某商贸大楼部分店面网络接入场景

（备注：具体的设备连接接口信息可根据实际情况决定）

【设备清单】

二层交换机（两台），三层交换机（1台），网线（若干），PC（若干）。

任务 ㉑ 使用 RLDP 技术快速检测以太网链路故障

【关键技术】

RLDP 是一种用于快速检测以太网链路故障的链路协议。

早期的以太网链路检测机制只是利用物理连接的状态，通过物理层的自动协商来检测链路的连通性。但是，这种检测机制存在一定的局限性，在一些情况下无法为用户提供可靠的链路检测信息。例如，当光纤接口上的光纤接收线对接错时，由于光纤转换器的存在，会导致设备对端接口物理层是在线的，但实际对应的二层链路却无法通信；又如，两台以太网设备之间架设着一个中间网络，存在网络传输中继设备，如果这台中继设备出现故障，则会造成同样的问题。

针对以上种种网络物理层链路的情况，通过在接入交换机的接入端口上开启 RLDP 快速检测以太网链路故障，可以实现二层网络故障快速检测。此外，在接入交换机上实施 RLDP 技术检测链路故障时，还需要开启 BPDU Guard 功能。BPDU Guard 保护可以有效防护 BPDU 消息传输，如果该接入端口配置了 BPDU Guard 功能，则一旦该端口收到了 BPDU 报文，就进入 Error-disabled 状态，无法转发数据。

二层交换网络中的环路有如下几种常见的情形。

情形一：三角环路（两台核心交换机连接在同一台接入设备上）。对于这种环路情形，可以通过在三台交换机之间运行 MSTP 来实现环路切换与线路冗余，如图 21-2 所示。

情形二：核心交换机两根线都连接到同一台接入交换机上。对于这种环路情形，可以通过在核心交换机和接入交换机上配置二层链路聚合来实现链路冗余，如图 21-3 所示。

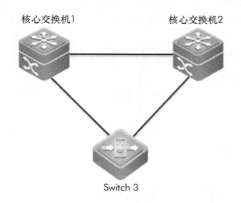

图 21-2 三角环路

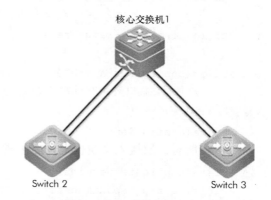

图 21-3 通过配置二层链路聚合实现链路冗余

情形三：同一根线的头尾两端插到同一台交换机的两个端口上，即设备自环。要防止这种网络环路，可以启用 RLDP 或 BPDU Guard 功能，如图 21-4 所示。

情形四：接入交换机连接集线器，集线器上的两个端口产生环路。对于这种网络场景，建议通过开启 RLDP 功能来检测，如图 21-5 所示。

由于集线器会过滤 BPDU 报文，不能使用 BPDU Guard 功能实现 BPDU 防护，因为接口上发出的目的 MAC 地址是 01-80-C2-00-00-00 的 BPDU 报文，市场上的集线器（如 TP-Link 或 D-Link 等）都会丢弃该类报文，无法转发 BPDU 报文。所以，即使集线器出现环路，也不会把 BPDU 报文转发给交换机，交换机也就无法把该连接的接口置为 Error-disabled 状态。

多层交换技术（实践篇）

图 21-4　设备自环

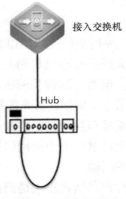

图 21-5　集线器自环

【实施步骤】

（1）在所有二层接入交换机上开启 RLDP 功能。

```
Ruijie#configure terminal
Ruijie(config)#rldp enable      // 全局开启 RLDP 功能

Ruijie(config)#interface range GigabitEthernet 0/1-24
            // 在下连计算机或集线器接口上开启，不要在接入交换机上连接口开启
Ruijie(config-if-range)#rldp port loop-detect shutdown-port
                    // 在接口上开启 RLDP 功能，如果检测出环路则 shutDown 该端口
Ruijie(config-if-range)#exit

Ruijie(config)#errdisable recovery interval 300
        // 如果该接口被 RLDP 检测到环路或链路故障，则 shutDown 该接口，300s 后会自动恢复，
并重新检测是否有环路
Ruijie(config)#end
```

（2）在所有二层接入交换机上开启生成树的保护功能。

```
Ruijie#configure terminal
Ruijie(config)#spanning-tree    //开启生成树协议，默认开启 MSTP

Ruijie(config)#interface range GigabitEthernet 0/1-24
Ruijie(config-if-range)#spanning-tree bpduguard enable
                    // 开启 BPDU Guard 保护功能
Ruijie(config-if-range)#spanning-tree portfast
                    // 在接入口上设置快速接口保护功能


Ruijie(config)#interface GigabitEthernet 0/25      //打开连接汇聚交换机的接口
Ruijie(config-if-GigabitEthernet 0/25)#spanning-tree bpdufilter enable
                    //在上连接口上开启 BPDUFilter 保护功能
```

任务 ㉑ 使用 RLDP 技术快速检测以太网链路故障

```
Ruijie(config-if-GigabitEthernet 0/25)#exit

Ruijie(config)#errdisable recovery interval 300
//如果该接口被 RLDP 检测到环路或链路故障，则 shutDown 该接口，300s 后自动恢复，并重新
检测是否有环路
Ruijie(config)#end
```

建议在下连接口上开启 RLDP 功能的同时，开启 BPDU Guard+Portfast 功能（注意，要使该功能生效，必须先开启生成树协议）。

如果网络中没有运行生成树协议（如单核心网络环境），则可以在接入交换机上开启生成树协议，同时，上连接口开启 BPDUFilter 功能，防止 BPDU 报文发送到核心交换机，影响整个网络。

【测试验证】

在设备上，使用"show rldp"命令查看 RLDP 的状态，如图 21-6 所示。

图 21-6　查看 RLDP 的状态

任务 22 实施环形 VSU 技术

【任务描述】

某网络中心为了提升网络的可靠性，在核心网络中部署了多台三层交换机互连形成冗余和备份，通过"MSTP+VRRP"技术实现核心网络稳健运行。但在传统技术支持下的冗余网络架构，不仅会提高网络设计和操作的复杂性，大量的备份链路还会降低网络中带宽资源的利用率。

随着网络虚拟化技术的广泛应用，虚拟交换单元（Virtual Switching Unit，VSU）网络系统虚拟化技术支持将多台三层交换机设备组合成单一虚拟设备，不仅可以简化网络拓扑，还可以降低网络的管理维护成本，缩短应用恢复的时间和业务中断的时间，提高网络中资源的利用率。

【任务目标】

实施环形 VSU 技术。

【组网拓扑】

图 22-1 所示网络拓扑为某网络中心核心网络场景，实施环形的 VSU 网络系统虚拟化技术，提升网络的可靠性。

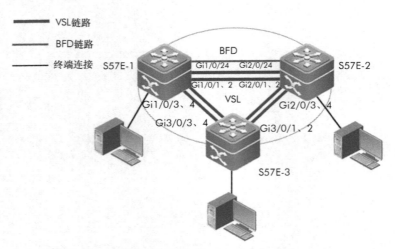

图 22-1 某网络中心核心网络场景

（备注：具体的设备连接接口信息可根据实际情况决定）

任务 22　实施环形 VSU 技术

需要注意的事项包括以下几点。

（1）使用 3 台三层交换机组建网络中心环形网络，其中，三层交换机 S57E-1、S57E-2 和 S57E-3 组成虚拟设备 VSU（域标识为 1），配置三层交换机 S57E-1 的优先级为 200，三层交换机 S57E-2 的优先级为 150，S57E-3 的优先级为 100。

如图 22-1 所示，使用 VSL 链路采用两两互连的方式连接。为了保证稳定性，建议两台设备互连的 VSL 至少为两条。如果条件有限，一条也可以做 VSU。

（2）VSU 实施完成后，基于 BFD 检测双主机，Gi1/0/24 和 Gi2/0/24 是一对 BFD 心跳接口，在配置过程中，需要随时测试 VSL 链路断开后的情况。

【设备清单】

三层交换机（3 台），万兆模块（若干，可选），万兆线缆（若干，可选），网线（若干），PC（若干）。

【关键技术】

VSU 是一种网络系统虚拟化技术，支持将多台设备组合成单一的虚拟设备，如图 22-1 所示。其中，接入、汇聚、核心层设备都可以组成 VSU 系统，形成整网端到端的 VSU 组网方案。和传统的组网方式相比，这种组网方式可以简化网络拓扑，降低网络的管理维护成本，缩短应用恢复的时间和业务中断的时间，提高网络资源的利用率。

在 VSU 实施的过程中，关于多台设备之间的连接，选择采用环形还是线形方式构建 VSU 防止双主机问题，主要基于以下几方面的考虑。

1. 基于 BFD 检测

一般情况下，无论是线形还是环形，建议首尾两台设备互连构成 BFD（环形 VSU 使用任意两台互连）检测系统。在特殊情况下，如果要完全防止双主机的产生，需要交换机之间两两互连来实现，也就是说，如果有 n 台交换机，需要使用 $n*(n-1)/2$ 条 BFD 线路。

如图 22-2 所示，采用 3 台 S5700E 设备做 VSU，如果要完全防止双主机问题，则需要使用一条 BFD 线路检测双主机系统。

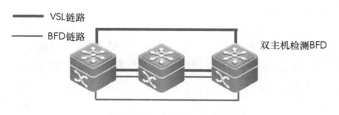

图 22-2　基于 BFD 检测

2. 基于聚合口检测

在实施多台设备之间组建的 VSU 技术时，需要实施防止双主机检测技术，最好的方式是采用聚合检测，这样不用像 BFD 那样需要多条 BFD 线路。聚合检测只需要 n 条线路即可，但前提是下连的接入交换机是同一厂商的设备，这样能够保证厂商的私有报文可以正常转发，如图 22-3 所示。

多层交换技术（实践篇）

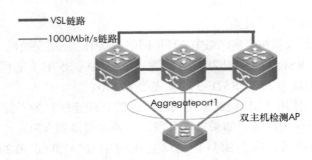

图 22-3　基于聚合口检测

【实施步骤】

（1）在三层交换机 S57E-1 上，配置 VSU 域标识、设备编号和优先级。

```
Ruijie#configure terminal
Ruijie(config)#hostname S57E-1
S57E-1(config)#switch virtual domain 1           // 创建 VSU 域标识
S57E-1(config-vs-domain)#switch 1                // 创建设备编号
S57E-1(config-vs-domain)#switch 1 priority 200   // 配置设备优先级
S57E-1(config-vs-domain)#switch 1 description S57E-1   // 配置设备描述信息
S57E-1(config-vs-domain)#exit
```

（2）在三层交换机 S57E-1 上配置 VSL 链路（基于锐捷 RGNOS 11.X 软件平台）。

（备注：部分版本及部分设备只能创建一条 VSL 链路，可使用 "vsl-port" 命令直接创建 1 个 VSL 链路组。）

```
S57E-1(config)#vsl-port 1                        // 创建 VSL 链路组 1
          // 或 S57E-1(config)# vsl-port         // 只能创建一个 VSL 链路组
S57E-1(config-vsl-ap-1)#port-member interface TenGigabitEthernet 0/1
                                 // 将 TE0/1 加入 VSL 组 1，也可以是吉比特电口
S57E-1(config-vsl-ap-1)#port-member interface TenGigabitEthernet 0/2
                                 // 将 TE0/2 加入 VSL 组 1，也可以是吉比特电口
S57E-1(config-vsl-ap-1)#exit

S57E-1(config)#vsl-port 2                        // 创建 VSL 链路组 2（可选）
          // 如果设备只能创建一个 VSL 链路组，则此处省略，以下 TE 加入上一个 VSL 链路组
S57E-1(config-vsl-ap-1)#port-member interface TenGigabitEthernet 0/3
                                 // 将 TE0/3 加入 VSL 组 2
S57E-1(config-vsl-ap-1)#port-member interface TenGigabitEthernet 0/4
                                 // 将 TE0/4 加入 VSL 组 2
S57E-1(config-vsl-ap-1)#exit
```

（3）在三层交换机 S57E-2 上配置 VSU 域标识、设备编号和优先级。

```
Ruijie#configure terminal
Ruijie(config)#hostname S57E-2
S57E-2(config)#switch virtual domain 1           // 创建 VSU 域标识
```

```
S57E-2(config-vs-domain)#switch 2                    // 创建设备编号
S57E-2(config-vs-domain)#switch 2 priority 150       // 配置设备的优先级
S57E-2(config-vs-domain)#switch 2 description S57E-2
                                                     // 配置设备描述信息
S57E-2(config-vs-domain)#exit
```

(4)在三层交换机 S57E-2 上配置 VSL 链路(基于锐捷 RGNOS 11.X 软件平台)。

(备注:部分版本及部分设备只能创建一条 VSL 链路,可以使用"vsl-port"命令直接创建 1 个 VSL 链路组。)

```
S57E-2(config)#vsl-port 1            // 创建 VSL 链路组 1
              // 或使用只能创建一个 VSL 链路组的命令: S57E-2(config)#vsl-port
S57E-2(config-vsl-ap-1)#port-member interface TenGigabitEthernet 0/1
                                     // 将 TE0/1 加入 VSL 组 1,也可以是吉比特电口
S57E-2(config-vsl-ap-1)#port-member interface TenGigabitEthernet 0/2
                                     // 将 TE0/2 加入 VSL 组 1,也可以是吉比特电口
S57E-2(config-vsl-ap-1)#exit

S57E-2(config)#vsl-port 2            // 创建 VSL 链路组 2(可选)
   // 如果设备只能创建一个 VSL 链路组,则此处省略,以下 TE 加入上一个 VSL 链路组
S57E-2(config-vsl-ap-1)#port-member interface TenGigabitEthernet 0/3
                                                     // 将 TE0/3 加入 VSL 组 2
S57E-2(config-vsl-ap-1)#port-member interface TenGigabitEthernet 0/4
                                                     // 将 TE0/4 加入 VSL 组 2
S57E-2(config-vsl-ap-1)#exit
```

(5)在三层交换机 S57E-3 上配置 VSU 域标识、设备编号和优先级。

```
Ruijie#configure terminal
Ruijie(config)#hostname S57E-3
S57E-3(config)#switch virtual domain 1               // 创建 VSU 域标识
S57E-3(config-vs-domain)#switch 3                    // 创建设备编号
S57E-3(config-vs-domain)#switch 3 priority 100       // 配置设备优先级
S57E-3(config-vs-domain)#switch 3 description S57E-3
                                                     // 配置设备描述信息
S57E-3(config-vs-domain)#exit
```

(6)在三层交换机 S57E-3 上配置 VSL 链路(基于锐捷 RGNOS 11.X 软件平台)。

(备注:部分版本及部分设备只能创建一条 VSL 链路,可以使用"vsl-port"命令直接创建 1 个 VSL 链路组。)

```
S57E-3(config)#vsl-port 1            // 创建 VSL 链路组 1
              // 或使用只能创建一个 VSL 链路组的命令: S57E-3(config)# vsl-port
S57E-3(config-vsl-ap-1)#port-member interface TenGigabitEthernet 0/1
                                     // 将 TE0/1 加入 VSL 组 1,也可以是吉比特电口
S57E-3(config-vsl-ap-1)#port-member interface TenGigabitEthernet 0/2
```

```
                                      // 将TE0/2加入VSL组1，也可以是吉比特电口
S57E-3(config-vsl-ap-1)#exit

S57E-3(config)#vsl-port 2         // 创建VSL链路组2（可选）
    // 如果设备只能创建一个VSL链路组，则此处省略，以下TE加入上一个VSL链路组
S57E-3(config-vsl-ap-1)#port-member interface TenGigabitEthernet 0/3
                                      // 将TE0/3加入VSL组2
S57E-3(config-vsl-ap-1)#port-member interface TenGigabitEthernet 0/4
                                      // 将TE0/4加入VSL组2
S57E-3(config-vsl-ap-1)#exit
```

（7）把三层交换机 S57E-1、S57E-2 和 S57E-3 交换机都转换为 VSU 模式。

```
S57E-1#switch convert mode virtual      // 将交换机S57E-1转换为VSU模式
Convert switch mode will automatically backup the "config.text" file and
then delete it, and reload the switch. Do you want to convert switch to
virtual mode?
                     [no/yes] y        // 输入y


S57E-2#switch convert mode virtual      // 将交换机S57E-2转换为VSU模式
Convert switch mode will automatically backup the "config.text" file
and then delete it, and reload the switch. Do you want to convert switch to
virtual mode?
                     [no/yes] y        // 输入y


S57E-3#switch convert mode virtual      // 将交换机S57E-3转换为VSU模式
Convert switch mode will automatically backup the "config.text" file
and then delete it, and reload the switch. Do you want to convert switch to
virtual mode?
                     [no/yes] y        // 输入y
```

（8）交换机会重启，进行 VSU 选举，重启时间会比较长，需要耐心等待。等待 VSU 建立成功后，进行 BFD 检测配置（基于锐捷 RGNOS 11.X 软件平台）。

```
Ruijie#configure terminal
Ruijie(config)#interface GigabitEthernet 1/0/24
                                      // 第一台VSU设备的第24个接口
Ruijie(config-if-GigabitEthernet 1/0/24)#no switchport
              // 只需要在BFD接口上输入switchport命令，无需其他配置
Ruijie(config-if-GigabitEthernet 1/0/24)#exit

Ruijie(config)#interface GigabitEthernet 2/0/24
```

```
                                              // 第二台 VSU 设备的第 24 个接口
Ruijie(config-if-GigabitEthernet 2/0/24)#no switchport
                             // 只需要在 BFD 接口上输入 switchport 命令，无需其他配置
Ruijie(config-if-GigabitEthernet 2/0/24)#exit

Ruijie (config)#switch virtual domain 1      //进入 VSU 参数配置模式
Ruijie(config-vs-domain)#dual-active detection bfd
                                          //打开 BFD 开关，其默认是关闭的
Ruijie(config-vs-domain)#dual-active bfd  interface Gi 1/0/24
                                                    // 配置一对 BFD 检测接口
Ruijie(config-vs-domain)#dual-active bfd  interface Gi 2/0/24
Ruijie(config-vs-domain)#dual-active exclude interface  ten1/1/2
        // 指定例外接口，如果上连路由接口保留，则出现双主机时可以 telnet（可选配置）
/* 如果拓扑中没有上连路由接口，配置中又没有配置路由接口，那么这条命令会提示无法配置 */
Ruijie(config-vs-domain)#dual-active exclude interface  ten2/1/2
                                                   // 指定例外接口（可选配置）
/* 如果拓扑中没有上连路由接口，配置中又没有配置路由接口，那么这条命令会提示无法配置 */
```

【测试验证】

备注：对于不同版本、不同型号的设备，测试结果会稍有不同。

（1）查看 VSU 信息，显示 VSU 的三种状态，即 ACTIVE、STANDBY、CANDIDATE，如图 22-4 所示。

```
Ruijie#show switch virtual
Switch_id   Domain_id   Priority   Status   Role        Description
-------------------------------------------------------------------
1(1)        1(1)        200(200)   OK       ACTIVE      S57E-1
2(2)        1(1)        150(150)   OK       STANDBY     S57E-2
3(3)        1(1)        100(100)   OK       CANDIDATE   S57E-3
```

图 22-4 查看 VSU 信息

（2）查看三层交换机 S57E-1 的 VSU 配置信息，如图 22-5 所示。

（3）查看三层交换机 S57E-2 的 VSU 配置信息，如图 22-6 所示。

```
Ruijie#show switch virtual config
switch_id: 1 (mac: 1414.4b5b.ed8e)
!
switch virtual domain 1
!
switch 1
switch 1 priority 200
switch 1 description S57E-1
!
vsl-aggregateport 1
port-member interface GigabitEthernet 0/1 copper
port-member interface GigabitEthernet 0/2 copper
vsl-aggregateport 2
port-member interface GigabitEthernet 0/3 copper
port-member interface GigabitEthernet 0/4 copper
!
```

图 22-5 查看 S57E-1 设备的 VSU 配置信息

```
switch_id: 2 (mac: 1414.4b60.422a)
!
switch virtual domain 1
!
switch 2
switch 2 priority 150
switch 2 description S57E-2
!
vsl-aggregateport 1
port-member interface GigabitEthernet 0/1 copper
port-member interface GigabitEthernet 0/2 copper
vsl-aggregateport 2
port-member interface GigabitEthernet 0/3 copper
port-member interface GigabitEthernet 0/4 copper
```

图 22-6 查看 S57E-2 设备的 VSU 配置信息

（4）查看三层交换机 S57E-3 的 VSU 配置信息，如图 22-7 所示。

（5）查看 VSU 的 BFD 检测信息，如图 22-8 所示。

```
switch_id: 3 (mac: 1414.4b76.7a16)
!
switch virtual domain 1
!
switch 3
switch 3 priority 100
switch 3 description S57E-3
!
vsl-aggregateport 1
port-member interface GigabitEthernet 0/1 copper
port-member interface GigabitEthernet 0/2 copper
vsl-aggregateport 2
port-member interface GigabitEthernet 0/3 copper
port-member interface GigabitEthernet 0/4 copper
!
```

图 22-7　查看 S57E-3 设备的 VSU 配置信息

```
Ruijie#show switch virtual dual-active summary
BFD dual-active detection enabled: Yes  BFD状态
Aggregateport dual-active detection enabled: NO
Interfaces excluded from shutdown in recovery mode:
In dual-active recovery mode: NO
```

图 22-8　查看 VSU 的 BFD 检测信息

（6）查看 VSL 信息，如图 22-9 所示。

```
Ruijie#show switch virtual link
VSL-AP   State   Peer-VSL   Rx                  Tx                  Uptime
--------------------------------------------------------------------------------
1/1      UP      2/1        56404               26749               0d,0h,5m
1/2      UP      3/1        66471               24657               0d,0h,58m
2/1      UP      1/1        26750               56404               0d,0h,5m
2/2      UP      3/2        65131               62526               0d,0h,58m
3/1      UP      1/2        24687               66553               0d,0h,58m
3/2      UP      2/2        62635               65217               0d,0h,58m
```

图 22-9　查看 VSL 信息

（7）查看 VSU 拓扑信息，如图 22-10 所示。

```
Ruijie#show switch virtual topology
Introduction: '[num]' means switch num, '(num/num)' means vsl-aggregatepo

Ring Topology:
[1](1/1)---(2/1)[2](2/2)---(3/2)[3](3/1)---(1/2)[1]

Switch[1]: ACTIVE, MAC: 1414.4b60.422a, Description: S57E-1
Switch[2]: STANDBY, MAC: 1414.4b1b.546c, Description: S57E-2
Switch[3]: CANDIDATE, MAC: 1414.4b76.7a16, Description: S57E-3
```

图 22-10　查看 VSU 拓扑信息

（8）将三层交换机 S57E-1 和 S57E-2 的两条 VSL 断开并进行观察，此时 VSU 拓扑成为线形，如图 22-11 所示。

```
Ruijie#show switch virtual
Switch_id     Domain_id     Priority     Status     Role          Description
---------------------------------------------------------------------------------
1(1)          1(1)          200(200)     OK         ACTIVE        S57E-1
2(2)          1(1)          150(150)     OK         STANDBY       S57E-2
3(3)          1(1)          100(100)     OK         CANDIDATE     S57E-3
Ruijie#show switch virtual topology
Introduction: '[num]' means switch num, '(num/num)' means vsl-aggregate
port num.

Chain Topology:
[1](1/2)---(3/1)[3](3/2)---(2/2)[2]

Switch[1]: ACTIVE, MAC: 1414.4b60.422a, Description: S57E-1
Switch[2]: STANDBY, MAC: 1414.4b1b.546c, Description: S57E-2
Switch[3]: CANDIDATE, MAC: 1414.4b76.7a16, Description: S57E-3
```

图 22-11　VSU 拓扑成为线形

（9）将 S57E-1 和 S57E-3 的两条 VSL 断开并进行观察，现在 S57E-1 单独与 S57E-2 和 S57E-3 分离，即 VSU 拓扑分离，如图 22-12 所示。

```
Ruijie#show switch virtual
Switch_id     Domain_id     Priority     Status     Role          Description
---------------------------------------------------------------------------------
1(1)          1(1)          200(200)     OK         ACTIVE        S57E-1
```

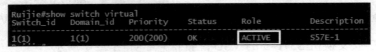

图 22-12　VSU 拓扑分离

（10）观察 S57E-2、S57E-3，进入 recovery 模式，如图 22-13 所示。

图 22-13　查看 recovery 模式

（11）把线缆重新插好，S57E-2、S57E-3 自动重启并加入 VSU，如图 22-14 所示。

图 22-14　S57E-2、S57E-3 自动重启并加入 VSU

任务 ㉓ 配置线形 VSU 技术

【任务描述】

某学院的网络中心为了提高网络的可靠性,通过多台骨干交换机互连形成冗余和备份,使用"MSTP+VRRP"技术实现核心网络的稳定性。但在传统技术支持下的冗余网络架构,不仅会提高网络设计和操作的复杂性,大量备份链路还会降低网络中带宽资源的利用率。

随着网络虚拟化技术的到来,使用 VSU 网络系统虚拟化技术,不仅可以简化网络拓扑,还可以降低网络的管理维护成本,缩短应用恢复的时间和业务中断的时间,提高网络资源的利用率。

【任务目标】

配置线形 VSU 技术。

【组网拓扑】

图 23-1 所示网络拓扑为某学院的网络中心网络场景,其实施了线形的 VSU 网络虚拟化技术,提升了网络的可靠性。

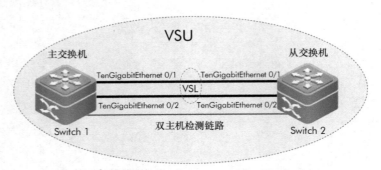

图 23-1 某学院的网络中心核心网络场景
(备注:具体的设备连接接口信息可根据实际情况决定)

【设备清单】

三层交换机(两台),万兆模块(若干,可选),万兆线缆(若干,可选),网线(若干),PC(若干)。

【关键技术】

使用 VSU 虚拟化技术可以实现网络中心核心交换机虚拟化连接,通过使用"VSU + BFD"

任务 23　配置线形 VSU 技术

技术可以实现核心网络的虚拟化保障。其中，VSU 提供虚拟化技术，使用 BFD 实施双主机检测，两项技术有机结合，防止 VSU 虚拟化中心跳线损坏后出现双主机的问题，如图 23-2 所示。

图 23-2　VSU 虚拟化连接

当 VSU 中的虚拟化的心跳线掉线后，两台交换机都会认为自己为 VSU 主机系统，导致两台 VSU 交换机上的三层地址出现冲突，使网络不稳定。使用 BFD 双主机检测技术，一旦 VSU 中互连的心跳有问题，BFD 会只让一台 VSU 主机存活，另一台主机系统除心跳接口及排除接口外，其他接口全部下线，从而避免 VSU 双主机现象的出现。

【实施步骤】

（1）在三层交换机 Switch 1 上配置 VSU 域标识、设备编号和优先级。

```
Ruijie#configure terminal
Ruijie(config)#hostname Switch 1
Switch 1#configure terminal
Switch 1(config)#switch virtual domain 1              // 创建 VSU 域标识
Switch 1(config-vs-domain)#switch 1                   // 创建设备编号
Switch 1(config-vs-domain)#switch 1 priority 200      // 配置设备优先级
Switch 1(config-vs-domain)#switch 1 description S57E-1 // 配置设备描述信息
Switch 1(config-vs-domain)#exit
```

（2）在三层交换机 Switch 1 上配置 VSL 链路（基于锐捷 RGNOS 11.X 软件平台）。

```
Switch 1(config)#
Switch 1(config)#vsl-port                             // 创建 VSL 链路组
Switch 1(config-vsl-ap-1)#port-member interface TenGigabitEthernet 0/1
                                       // 将 TE0/1 加入 VSL 组，也可使用吉比特电口
Switch 1(config-vsl-ap-1)#port-member interface TenGigabitEthernet 0/2
                                       // 将 TE0/2 加入 VSL 组，也可使用吉比特电口
// 部分交换机上可以直接使用吉比特端口启用 VSL 链路，和该设备的操作系统版本有关
```

（3）在三层交换机 Switch 2 上配置 VSU 域标识、设备编号和优先级。

```
Ruijie#configure terminal
Ruijie(config)#hostname Switch 2
Switch 2(config)#switch virtual domain 1              // 创建 VSU 域标识
```

125

```
Switch 2(config-vs-domain)#switch 2                        // 创建设备编号
Switch 2(config-vs-domain)#switch 2 priority 150           // 配置设备优先级
Switch 2(config-vs-domain)#switch 2 description S57E-2     // 配置设备描述信息
Switch 2(config-vs-domain)#exit
```

（4）在三层交换机 Switch 2 上配置 VSL 链路（基于锐捷 RGNOS 11.X 软件平台）。

```
Switch 2(config)#vsl-port                                  // 创建 VSL 链路组
Switch 2(config-vsl-ap-1)#port-member interface TenGigabitEthernet 0/1
                                                           // 将 TE0/1 加入 VSL 组
Switch 2(config-vsl-ap-1)#port-member interface TenGigabitEthernet 0/2
                                                           // 将 TE0/2 加入 VSL 组
Switch 2(config-vsl-ap-1)#exit
// 部分交换机上可以直接使用吉比特端口启用 VSL 链路，和该设备的操作系统版本有关
```

（5）将三层 Switch 1、Switch 2 转换为 VSU 模式。

```
Switch 1#switch convert mode virtual      // 将交换机 Switch 1 转换为 VSU 模式
Convert switch mode will automatically backup the "config.text" file and
then delete it, and reload the switch. Do you want to convert switch to
virtual mode?
                         [no/yes]   y      // 输入 y

Switch 2#switch convert mode virtual      // 将交换机 Switch 2 转换为 VSU 模式
Convert switch mode will automatically backup the "config.text" file and
then delete it, and reload the switch. Do you want to convert switch to
virtual mode?
                         [no/yes]   y      // 输入 y
```

（6）交换机重启，进行 VSU 的选举，重启时间会比较长，需要耐心等待。等待 VSU 建立成功后，在主交换机上查看 VLAN 信息、拓扑信息。

```
Ruijie-STANDBY#show vlan
…
Ruijie-STANDBY#show switch virtual topology
…
```

（7）等待 VSU 建立成功后，进行 BFD 配置（基于锐捷 RGNOS 11.X 软件平台）。

```
Ruijie-STANDBY#
Ruijie-STANDBY#configure terminal
Ruijie-STANDBY(config)#interface GigabitEthernet 1/0/24
                                    // 第一台 VSU 设备 Gi0/24 接口作为双主机检测接口
Ruijie-STANDBY(config-if-GigabitEthernet 1/0/24)#no switchport
                                                  // BFD 检测在三层口上完成
Ruijie-STANDBY#(config-if-GigabitEthernet 1/0/24)#exit

Ruijie-STANDBY(config)#interface GigabitEthernet 2/0/24
```

任务 ㉓ 配置线形 VSU 技术

```
                                      // 第二台 VSU 设备 Gi0/24 接口作为双主机检测接口
Ruijie-STANDBY(config-if-GigabitEthernet 2/0/24)#no switchport
                                                      // BFD 检测需在三层接口上完成
Ruijie-STANDBY(config-if-GigabitEthernet 2/0/24)#exit

Ruijie -STANDBY(config)#
Ruijie-STANDBY(config)# switch virtual domain 1
                            // 进入 VSU 参数配置模式，进行 BFD 配置
Ruijie-STANDBY(config-vs-domain)# dual-active detection bfd
                                          // 打开 BFD 开关，其默认是关闭的
Ruijie-STANDBY(config-vs-domain)# dual-active bfd interface Gi 1/0/24
                                                  // 配置一对 BFD 检测接口
Ruijie-STANDBY(config-vs-domain)# dual-active bfd interface Gi 2/0/24
Ruijie-STANDBY(config-vs-domain)# dual-active exclude interface ten1/1/2
          // 指定例外端口，如果上连路由口保留，则出现双主机时可以远程登录（可选）
/* 如果拓扑中没有上连路由口，又没有配置路由口，那么这条命令会提示无法配置 */
Ruijie-STANDBY(config-vs-domain)# dual-active exclude interfaceTen2/1/2
                                                       // 指定例外端口（可选）
/* 如果拓扑中没有上连路由口，又没有配置路由口，那么这条命令会提示无法配置 */
```

【测试验证】

备注：对于不同版本、不同型号的设备，测试结果会稍有不同。

（1）查看 VSU 信息，显示 VSU 两种状态，即 Active、Standby。

```
Ruijie-STANDBY#show switch virtual
Topology status: Converge
Switch_id  Domain_id  Priority   Position  Status  Mode     Role     Description
1(1)       1(1)       200(200)   REMOTE    OK      Virtual  Active   S57E-1
2(2)       1(1)       150(150)   LOCAL     OK      Virtual  Standby  S57E-2
```

（2）查看 Switch 1 的 VSU 配置信息。

```
Ruijie-STANDBY#show switch virtual config
Switch_id: 1 (mac: 8005.8847.2da9)
switch virtual domain 1
switch 1
switch 1 priority 200
vsl-aggregateport 1
port-member interface TenGigabitEthernet 1/0/1
port-member interface TenGigabitEthernet 1/0/2
switch convert mode virtual

Switch_id: 2 (mac: 8005.8847.2ea5)
```

多层交换技术（实践篇）

```
switch virtual domain 1
switch 2
switch 2 priority 150
vsl-aggregateport 1
port-member interface TenGigabitEthernet 2/0/1
port-member interface TenGigabitEthernet 2/0/2
switch convert mode virtual
```

（3）查看 Switch 1 的虚拟链路配置信息。

```
Ruijie-STANDBY#show switch virtual link
VSL-AP      State     Peer-VSL    Rx packets      Tx packets
2/1         UP        1/1         3976            4029
1/1         UP        2/1         4003            3998
```

任务 24 配置 VSU 综合应用案例

【任务描述】

某企业网络中心为实现高可靠性，拟在网络中心的核心网中实现冗余和备份，保障高带宽的传输。为提升网络中心网络的稳定性，拟在网络中心安装两台万兆核心交换机产品 S8606-1 与 S8606-2，配置基于 VSL 双链路模式，组建网络中心网络 VSU 系统。

【任务目标】

配置基于 VSL 双链路的 VSU 综合应用。

【组网拓扑】

图 24-1 所示网络拓扑为某企业网络中心场景，配置基于 VSL 双链路的 VSU 虚拟化技术，保障网络中心的高可靠性。

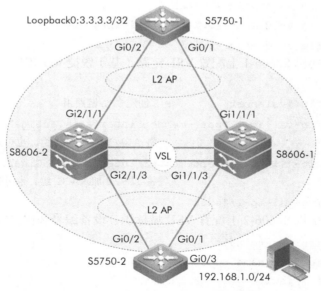

图 24-1 某企业网络中心组网场景

（备注：具体的设备连接接口信息可根据实际情况决定）

【关键技术】

在配置本任务的基于 VSL 链路的 VSU 虚拟化技术时，需要注意如下事项。

VSU 通过核心交换机 S8606-1 上的 Gi1/1/1 接口和 S8606-2 上的 Gi2/1/1 接口，建立跨

多层交换技术（实践篇）

机二层聚合口；同时，通过 S8606-1 上的 Gi1/1/3 接口与 S8606-2 上的 Gi2/1/3 接口，建立跨交换机框架的二层聚合口。

S5750-1 交换机通过 Gi0/1 和 Gi0/2 接口与两台核心交换机 S8606 互连，并实施 VSU 虚拟化技术，建立二层聚合链路。S5750-2 交换机通过 Gi0/1 和 Gi0/2 接口与核心交换机 S8606 互连，并实施 VSU 虚拟化技术，建立二层聚合链路。

交换机 S5750-1、S5750-2 与 VSU 系统之间通过 OSPF 协议通告路由信息。其中，交换机 S5750-1 通过 OSPF 通告 Loopback0 地址为 3.3.3.3/32；交换机 S5750-2 通过 OSPF 协议通告子网地址为 192.168.1.0/24。

在全网的 VSU 系统中，启用 BFD 双主机检测功能。

【设备清单】

三层交换机（4 台），万兆模块（若干，可选），万兆线缆（若干，可选），网线（若干），PC（若干）。

【实施步骤】

（1）在三层交换机 S8606-1 上配置 VSU 域标识、设备编号和优先级。

```
Ruijie#configure terminal
Ruijie(config)#hostname S8606-1
S8606-1(config)#switch virtual domain 1              // 创建 VSU 域标识
S8606-1(config-vs-domain)#switch 1                   // 创建设备编号
S8606-1(config-vs-domain)#switch 1 priority 200      // 配置设备优先级
S8606-1(config-vs-domain)#exit
```

（2）在三层交换机 S8606-1 上配置 VSL 链路（基于锐捷 RGNOS 11.X 软件平台）。

```
S8606-1(config)#
S8606-1(config)#vsl-port                             // 创建 VSL 链路组
S8606-1(config-vsl-ap-1)#port-member interface TenGigabitEthernet 0/1
                                                     // 将 TE0/1 加入 VSL 组，也可选择吉比特电口
S8606-1(config-vsl-ap-1)#port-member interface TenGigabitEthernet 0/2
                                                     // 将 TE0/2 加入 VSL 组，也可选择吉比特电口
S8606-1(config-vsl-ap-1)#exit
```

（3）在三层交换机 S8606-2 上配置 VSU 域标识、设备编号和优先级。

```
Ruijie#configure terminal
Ruijie(config)#hostname S8606-2
S8606-2(config)#switch virtual domain 1              // 创建 VSU 域标识
S8606-2(config-vs-domain)#switch 2                   // 创建设备编号
S8606-2(config-vs-domain)#switch 2 priority 150      // 配置设备优先级
S8606-2(config-vs-domain)#exit
```

（4）在三层交换机 S8606-2 上配置 VSL 链路（基于锐捷 RGNOS 11.X 软件平台）。

```
S8606-2(config)#
S8606-2(config)#vsl-port                             // 创建 VSL 链路组
```

任务 ㉔ 配置 VSU 综合应用案例

```
S8606-2(config-vsl-ap-1)#port-member interface TenGigabitEthernet 0/1
                        // 将 TE0/1 加入 VSL 组，也可选择吉比特电口
S8606-2(config-vsl-ap-1)#port-member interface TenGigabitEthernet 0/2
                        // 将 TE0/2 加入 VSL 组，也可选择吉比特电口
S8606-2(config-vsl-ap-1)#exit
```

（5）把三层交换机 S8606-1 和 S8606-2 从单机模式转换为 VSU 模式。

```
S8606-1#switch convert mode virtual      // 将交换机 S8606-1 转换为 VSU 模式
Convert switch mode will automatically backup the "config.text" file and
then delete it, and reload the switch. Do you want to convert switch to
virtual mode?
                        [no/yes]  y      // 输入 y
...
```

```
S8606-2#switch convert mode virtual      // 将交换机 S8606-2 转换为 VSU 模式
Convert switch mode will automatically backup the "config.text" file and
then delete it, and reload the switch. Do you want to convert switch to
virtual mode?
                        [no/yes]  y      // 输入 y
...
```

（6）等待 VSU 建立成功后，在激活的 VSU 系统上配置二层聚合口 AP。
① 创建聚合口 1。

```
Ruijie-Active(config)#
Ruijie-Active(config)#interface GigabitEthernet 1/1/1
Ruijie-Active(config-if-GigabitEthernet 1/1/1)#switchport
Ruijie-Active(config-if-GigabitEthernet 1/1/1)#port-group 1
                        // 打开接口并将其加入聚合口 1
Ruijie-Active(config-if-GigabitEthernet 1/1/1)#interface GigabitEthernet 2/1/1
Ruijie-Active(config-if-GigabitEthernet 2/1/1)#switchport
Ruijie-Active(config-if-GigabitEthernet 2/1/1)#port-group 1
                        // 打开接口并将其加入聚合口 1
Ruijie-Active(config-if-GigabitEthernet 2/1/1)#exit
Ruijie-Active(config)#
```

② 创建聚合口 2。

```
Ruijie-Active(config)#
Ruijie-Active(config)# interface GigabitEthernet 1/1/3
Ruijie-Active(config-if-GigabitEthernet 1/1/3)#switchport
Ruijie-Active(config-if-GigabitEthernet 1/1/3)#port-group 2
                        // 打开接口并将其加入聚合口 2
Ruijie-Active(config-if-GigabitEthernet 1/1/3)#interface GigabitEthernet
```

```
2/1/3
    Ruijie-Active(config-if-GigabitEthernet 2/1/3)#switchport
    Ruijie-Active(config-if-GigabitEthernet 2/1/3)#port-group 2
                                        // 打开接口并将其加入聚合口 2
    Ruijie-Active(config-if-GigabitEthernet 2/1/3)#exit

    Ruijie-Active(config)#
    Ruijie-Active(config)#interface AggregatePort 1      // 激活聚合口 1
    Ruijie-Active(config-if-AggregatePort 1)#exit
    Ruijie-Active(config)#interface AggregatePort 2      // 激活聚合口 2
    Ruijie-Active(config-if-AggregatePort 2)#exit
```

③ 配置创建完成的聚合口为 Trunk 模式。

```
    Ruijie-Active(config)#interface AggregatePort 1
    Ruijie-Active(config-if-AggregatePort 1)#switchport mode trunk
    Ruijie-Active(config-if-AggregatePort 1)#exit
    Ruijie-Active(config)#interface AggregatePort 2
    Ruijie-Active(config-if-AggregatePort 2)#switchport mode trunk
    Ruijie-Active(config-if-AggregatePort 2)#exit
```

④ 创建 VSU 系统上的 SVI 接口。

```
    Ruijie-Active(config)#vlan 101
    Ruijie-Active(config-vlan)#vlan 102
    Ruijie-Active(config-vlan)#exit
    Ruijie-Active(config)#
```

⑤ 给 SVI 接口配置地址。

```
    Ruijie-Active(config)#interface vlan 101
    Ruijie-Active(config-ip-vlan 101)#ip address 1.1.1.2 255.255.255.252
    Ruijie-Active(config-ip-vlan 101)#interface vlan 102
    Ruijie-Active(config-ip-vlan 102)#ip address 2.2.2.2 255.255.255.252
    Ruijie-Active(config-ip-vlan 102)#interface Loopback 0
    Ruijie-Active(config-if-Loopback 0)#ip address 11.11.11.11 255.255.255.255
    Ruijie-Active(config-if-Loopback 0)#exit
    Ruijie-Active(config)#
```

（7）启用 IP 的 BFD 双主机检测功能。

```
    Ruijie-Active(config)#
    Ruijie-Active(config)#interface GigabitEthernet 1/2/8
                              // 核心交换机 S8606-1 和 S8606-2 连接的接口
    Ruijie-Active(config-if-GigabitEthernet 1/2/8)#no switchport
    Ruijie-Active(config-if-GigabitEthernet 1/2/8)#no ip proxy-arp
    Ruijie-Active(config-if-GigabitEthernet 1/2/8)#ip address 200.200.200.1
255.255.255.0
```

任务 ㉔ 配置 VSU 综合应用案例

```
    Ruijie-Active(config-if-GigabitEthernet 1/2/8)#bfd interval 50 min_rx 50 multiplier 3
    Ruijie-Active(config-if-GigabitEthernet 1/2/8)#exit
                                        // 设置 BFD 心跳接口的 BFD 参数

    Ruijie-Active(config)#
    Ruijie-Active(config)#interface GigabitEthernet 2/2/8
                            // 核心交换机 S8606-1 和 S8606-2 连接的接口
    Ruijie-Active(config-if-GigabitEthernet 2/2/8)#no switchport
    Ruijie-Active(config-if-GigabitEthernet 2/2/8)#no ip proxy-arp
    Ruijie-Active(config-if-GigabitEthernet 2/2/8)#ip address 200.200.200.2 255.255.255.0
    Ruijie-Active(config-if-GigabitEthernet 2/2/8)#bfd interval 50 min_rx 50 multiplier 3
    Ruijie-Active(config-if-GigabitEthernet 2/2/8)#exit
    //设置 BFD 心跳接口的 BFD 参数

    Ruijie-Active(config)#
    Ruijie-Active(config)#switch virtual domain 1
    Ruijie-Active(config-vs-domain)#dual-Active detection bfd
                                        // 启用三层的 BFD 双主机检测功能
    Ruijie-Active(config-vs-domain)#dual-Active pair interface GigabitEthernet 1/2/8 interface GigabitEthernet 2/2/8 bfd  // 配置 BFD 心跳接口对
```

（8）在激活的 VSU 上启用 OSPF 协议，通告子网路由。

```
    Ruijie-Active(config)#
    Ruijie-Active(config)#router ospf
    Ruijie-Active(config-router)#router-id 11.11.11.11
    Ruijie-Active(config-router)#network 2.2.2.2 0.0.0.255 area 0
    Ruijie-Active(config-router)#network 1.1.1.2 0.0.0.255 area 0
    Ruijie-Active(config-router)#network 11.11.11.11 0.0.0.255 area 0
```

（9）在 S5750-1 上配置二层聚合口，启用 OSPF 协议，通告子网路由。

```
    Ruijie(config)#
    Ruijie(config)#interface rang Gi0/1-2
    Ruijie(config-if-range)#switchport
    Ruijie(config-if-range)#port-group 1
    Ruijie(config-if-range)#exit
    Ruijie(config)#interface AggregatePort 1
    Ruijie(config-if-AggregatePort 1)#switchport mode trunk

    Ruijie(config-if-range)#interface vlan 101
```

```
Ruijie(config-if-vlan 101)#ip address 1.1.1.1 255.255.255.252
Ruijie(config-if-vlan 101)#exit


Ruijie(config)#interface Loopback 0
Ruijie(config-if-Loopback 0)#ip address 3.3.3.3 255.255.255.255
Ruijie(config-if-Loopback 0)#no shutdown
Ruijie(config-if-Loopback 0)#exit


Ruijie(config)#router ospf
Ruijie(config-router)#network 1.1.1.1 0.0.0.255 are 0
Ruijie(config-router)#network 3.3.3.3 0.0.0.255 are 0
Ruijie(config-router)#exit
```

（10）在 S5750-2 上配置二层聚合口，启用 OSPF 协议，通告子网路由。

```
Ruijie(config)#
Ruijie(config)#interface rang Gi0/1-2
Ruijie(config-if-range)#switchport
Ruijie(config-if-range)#port-group 1
Ruijie(config-if-range)# exit

Ruijie(config)#interface Aggregateport 1
Ruijie(config-if-AggregatePort 1)#switchport mode trunk
Ruijie(config-if-AggregatePort 1)#exit

Ruijie(config)#interface vlan 102
Ruijie(config-if-vlan 101)#ip address 2.2.2.1 255.255.255.252
Ruijie(config-if-vlan 101)#exit

Ruijie(config)#interface Gi0/3       // 连接测试计算机的接口
Ruijie(config-if-FastEthernet 0/3)#no switchport
Ruijie(config-if-FastEthernet 0/3)#ip address 192.168.1.1 255.255.255.0
Ruijie(config-if-FastEthernet 0/3)#exit


Ruijie(config-router)#interface loopback0
Ruijie(config-if-Loopback 0)#ip address 4.4.4.4 255.255.255.0
Ruijie(config-if-Loopback 0)#no shutdown
Ruijie(config-if-Loopback 0)#end


Ruijie(config)#router ospf
Ruijie(config-router)#network 192.168.1.0 0.0.0.255 area 0
Ruijie(config-router)#network 2.2.2.1 0.0.0.255 area 0
```

任务 ㉔　配置 VSU 综合应用案例

```
Ruijie(config-router)#network 4.4.4.4 0.0.0.255 area 0
Ruijie(config-router)#exit
Ruijie(config)#
```

【测试验证】

备注：对于不同版本、不同型号的设备，测试结果会稍有不同。

（1）查看 VSU 的基本信息。

```
Ruijie-Active#show switch virtual
Switch mode                            : Virtual Switch
Virtual switch domain number   : 1
Local switch number                    : 1
Local switch operational role  : Active
Peer switch number                     : 2
Peer switch operational role   : Standby
```

```
Ruijie-Active#show switch virtual role
Switch     Switch    Status      Priority            Role
------------------------------------------------------------------
LOCAL      1         UP          200(200)            ACTIVE
REMOTE     2         UP          100(100)            STANDBY
In dual-Active recovery mode: No
```

（2）查看 VSL 的基本信息。

```
Ruijie-Active#show switch virtual link
VSL Status : UP
VSL Uptime : 1 day, 1 hours, 15 minutes
```

```
Ruijie-Active#show switch virtual link port
Port         State            Peer port       Rx          Tx
------------------------------------------------------------------
Te 1/3/1     OPERATIONAL      Te 2/3/1        100020      100033
Te 1/3/2     OPERATIONAL      Te 2/3/2        99889       100034
```

```
Ruijie-Active#show switch virtual link vslp statistics
Port Te 1/3/1
Received LMP request 100(98 ok, 2 error), LMP reply 100(95 ok, 1 error),
LMP keepalive 100(99 ok, 1 error)
Sent LMP request 100(99 ok, 1 fail), LMP reply 100(100 ok, 0 fail), LMP
keepalive 100(100 ok, 0 fail)
Port Te 1/3/2
Received LMP request 100(99 ok, 1 error), LMP reply 100(99 ok, 1 error),
```

```
LMP keepalive 100(99 ok, 1 error)
    Sent   LMP request 100(99 ok, 1 fail), LMP reply 100(99 ok, 1 fail), LMP
keepalive 100(99 ok, 1 fail)
    VSL
    Received RRP adv 100(100 ok, 0 error), RRP reply 100(100 ok, 0 error)
    Sent   RRP adv 100(98 ok, 2 fail), RRP reply 100(95 ok, 5 fail)
```

（3）查看双主机配置状态。

```
Ruijie-Active#show switch virtual dual-Active summary
BFD dual-Active detection enabled:   Yes
Interfaces excluded from shutdown in recovery mode:  Gi 1/1/3
In dual-Active recovery mode: No
```

（4）查看 IP BFD 双主机检测信息。

```
Ruijie-Active#show switch virtual dual-Active bfd
BFD dual-Active detection enabled: Yes
BFD dual-Active interface pairs configured:
Gi 1/2/8
Gi 2/2/8
```

任务 25 排除 VSU 故障

【任务描述】

某园区网络中心为实现可靠性，拟在网络中心的核心网中实现冗余和备份，保障高带宽的传输。为提升网络中心网络的稳定性，拟通过 VSU 虚拟化技术增强核心网络的可靠性。

【任务目标】

总结 VSU 施工经验，排除 VSU 项目故障。

【故障总结】

（1）锐捷网络有哪些设备能实现 VSU 功能？

目前，在锐捷网络的产品中，支持 VSU 功能的设备的型号如下。

10.X 盒式：S57E/P 系列、S6000 系列、S6220 10.X 平台系列、6200 系列（注：S29E 只支持堆叠）。

11.X 盒式：S2910 系列、S2910-POE 系列、S57XS-L 系列、S57H 系列、S6000E 系列、S6010 系列、S6220 11.X 平台系列及 S6220-H 系列。

10.X 机箱：S86 系列、S12000 系列。

11.X 机箱：S78C 系列、S78E 系列、S86E 系列、N18K 系列。

（2）部署 VSL 链路对接口有何要求？

目前，在部署 VSL 链路时，针对不同的设备，接口要求有所不同。

① 中低端设备：在 10.X 版本的盒式设备上，支持吉比特口做 VSU（注：吉比特电口和光口都支持，光模块必须是吉比特光模块）。在 11.X 版本的盒式设备上，只支持万兆口做 VSU（注：万兆光口做 VSU，光模块必须是万兆光模块）。

② 高端设备：在 10.X 版本的 S86 系列机箱式设备上，M8600-VSU-02XFP 线卡支持万兆口做 VSU；增强型业务线卡 EC 系列线卡的所有万兆口均可组建 VSU。

在 11.X 版本的机箱式设备上，支持万兆口或 40G 做 VSU（注意：万兆光口做 VSU，光模块必须是万兆光模块，40G 口不能拆分）。此外，组建 VSU 时要求系统版本一致。

（3）S6220 系列和 S6010 交换机在部署 VSL 时的限制条件及注意事项有哪些？

S6220 系列和 S6010 交换机默认的 4 个万兆口一组自动加入 VSL，如 9 号接口加入 VSL 后，10 号口、11 号口、12 号口均自动加入 VSL，不能作为普通接口使用，所以应提前规划接口的数量及用途，避免接口浪费。

锐捷 S6220-48XS 交换机的 Ten0/1-8 万兆口支持 FC，默认不支持作为 VSL 端口。如果需要让这些端口支持 VSL，需要使用 "convert interface slot 0 port x to ethernet" 命令进行转

换，转换后，保存配置并重启设备后才生效。

S6220 系列和 S6010 交换机的万兆端口如果降速为吉比特端口使用（如插入吉比特光模块），则此接口不能作为 VSL 端口（无法构建 VSU）使用。

S6220 系列和 S6010 交换机的 40G 接口能作为 VSL 端口使用，但是拆分为 10Gx4 模式后，这 4 个口都不能作为 VSL 端口使用。

（4）部署 BFD 链路检测对链路有什么要求？

VSU 系统支持使用 BFD 检测和 MAD 检测。

当配置完成的 VSL 链路上的所有物理链路异常断开时，网络中就会出现两台设备上的 VSU 配置完全相同的情况，导致网络不可用。配置 BFD 双主机检测功能，可以阻止以上异常故障出现。

与 VSL 链路相比，BFD 链路检测没有那些限制，只要链路两边能正常通信即可，即对速率、介质等没有要求，相当于普通的链路，万兆、吉比特、百兆都是可以的。

（5）VSU 配置基本步骤有哪些？

锐捷网络的设备上配置 VSU 包括以下三个基本步骤。

① 创建 VSU 域标识：switch virtual domain 1（注：组建 VSU 的域标识必须一致）。

② 配置设备编号：switch 1（注：设备编号不能一致，只在本机生效）。

③ 设置设备优先级：switch 1 priority 120（注：默认优先级为 100，配置为较高的优先级，VSU 建立成功后将会成为管理主机）。

④ 配置 VSL 链路：vsl-port（注：进入 VSL 链路组）。

此外，添加成员端口的命令为"port-member interface TenGigabitEthernet 2/1"。

（6）VSL 链路变更后，如何添加和删除接口？

在互连的交换机上激活 VSU 系统功能后，需要更改 VSL 链路的情况有如下几种。

一是由于网络改造的需要重新规划了接口用途，这种接口规划调整可能会导致 VSL 链路上的接口发生调整。

二是原有 VSL 链路或接口出现故障，需要更换接口。

针对以上情况，若要更改 VSL 链路，应进入 VSL 接口池，在接口池中增加或移除接口。变更操作在 VSU 模式下进行即可，不需要切换成单机模式，也不需要重启设备；所有配置只能在全局主设备上进行；VSL 链路两端的接口不用指定，但是接口数量要一致。

例如，Switch 1 与 Switch 2 组成 2 虚 1 的 VSU 系统，Switch 1 与 Switch 2 各给出两个接口组成 VSL 接口池，不允许出现 Switch 1 只给出 1 个接口、Switch 2 上给出两个接口，或者 Switch 1 给出两个接口、Switch 2 上给出 1 个接口的情况。

此外，在锐捷网络的 10.X 版本软件平台下，添加或删除 VSL 配置，可以增加或者删除某条 VSL 链路，但是当下不生效，需要重启设备才能生效，但其在高端箱式设备上会立即生效，具体代码示例如下。

```
Ruijie(config)#vsl-aggregateport 1/1        // 注意，VSU 模式下需要使用二维模式
Ruijie(config-vsl-ap-1/1)#port-member interface GigabitEthernet 0/20 fiber     //添加链路
Ruijie(config-vsl-ap-1/1)#no port-member interface GigabitEthernet 0/20
```

任务 ㉕ 排除 VSU 故障

```
fiber    // 删除链路
    The configuration of port GigabitEthernet 1/0/20 will be removed, do you
want to continue?
                                    [no/yes]  y
    % To update VSL firmwares, please save configuration and reload switch 1.
```
在锐捷网络的 11.X 版本软件平台下,需要通过如下配置实现 VSL 变更。

```
Ruijie#config
Ruijie(config)# vsl-port          // 进入 VSL-PORT 配置模式
Ruijie(config-vsl-port)# port-member interface Tengigabitethernet 1/2/3
// 增加 1/2/3 作为 VSL 接口
Ruijie(config-vsl-port)# no port-member interface Tengigabitethernet 1/2/2
// 将 1/2/2 从 VSL 接口中移除
```

(7) VSU 系统配置完成后存在哪四个相关的配置文件?

配置完成的 VSU 系统中存在四个相关的配置文件,它们分别是 config_vsu.dat、standalone.text、virtual_switch.text、config.text。通过使用 "dir" 命令,可以查看 Flash 中的这四个文件。

① config_vsu.dat:这个文件中存放的是关于组建 VSU 系统的配置,设备启动的时候会根据这个文件中的内容进行 VSU 系统协商与拓扑发现,可以使用 "Ruijie#show switch virtual config" 命令查看。

② standalone.text:这个文件中存放的是该物理设备曾经在单机模式下的运行配置信息,以便从 VSU 模式切换到单机模式的时候能够直接调用。当然,可以不加载这份曾经的单机模式下的配置,可以放弃,重新开始单机模式的相关配置。

③ virtual_switch.text:这个文件中存放的是该物理设备曾经在 VSU 模式下的运行配置信息,以便从单机模式切换到 VSU 模式的时候能够直接调用。当然,可以不加载这份曾经的 VSU 下的配置,可以放弃,重新开始 VSU 下的相关功能配置。

④ config.text:这个文件中存放的是不管是 VSU 还是单机模式下使用 "show run" 命令看到的配置,是系统目前工作状态下所加载的实际用于设备运行的配置。当然,如果设备工作在单机模式下,则其内容和 standalone.text 一致;如果设备工作在 VSU 模式下,则其内容和 virtual_switch.text 一致;如果在模式切换的时候放弃了文件直接替换的建议,那么应重新开始配置,它们之间的关系可能就不一致了。

了解了上述四个文件的作用后,如果出现如下问题,可以尝试使用对应的解决方案。

首先,如果发现执行 "Ruijie#switch convert mode standalone /virtual" 命令,不能切换单机/VSU 模式的时候,可以尝试直接删除 config_vsu.dat 文件和 config.text 文件,再进行模式切换。

其次,如果发现切换到单机模式后还能看到当初在 VSU 模式下的残留配置,或者是切换到 VSU 模式后还能看到当初在单机模式下的配置,则可以尝试直接删除 standalone.text 文件、virtual_switch.text 文件和 config.text 文件。这样,设备模式切换后就会呈空配置,即可重新开始新的配置。

多层交换技术（实践篇）

另外，如果希望模式切换只是临时测试使用，稍后可能还会重新切换回来，例如，当前是 VSU 模式，需要切换到单机模式下进行软件升级，启动成功后要马上切换回 VSU 模式运行，希望这个 VSU 模式下的配置不会丢失或修改，当重新切换回来的时候能够保持原来的运行状态，则推荐使用如下方法。

a. 确认当前的运行配置没有问题，执行 write 操作，将配置写入 virtual_switch.text 文件与 config.text 文件。

b. 执行模式切换，从 VSU 模式切换为单机模式。

c. 执行单机模式到 VSU 模式的切换。

在切换的过程中，使用"Switch 2# switch convert mode virtual"命令即可转换为 VSU 模式，在出现以下提示信息后，必须选择 yes，才可以将原来的 VSU 配置加载为设备运行的配置。

```
Are you sure to convert switch to virtual mode[yes/no]: yes
Do you want to recovery"config.text"from"virtual_switch.text" [yes/no]: yes
```

（8）在 VSU 环境下，系统升级和单机设备升级有什么不同？需要注意什么事项？

在 VSU 环境下升级与普通单机设备升级方式一致。

升级操作需要在主设备上进行，会自动同步备份设备完成系统升级。需要注意的是，系统升级过程中，需要详细阅读版本发行说明，确认硬件适用情况，且在升级过程中设备不能断电。

（9）安装 10.X 版本和 11.X 版本软件平台的设备，组建 VSU 时，其配置有什么区别？

① 在组建 VSU 时，配置 VSL 链路的命令不同。

在 10.X 版本软件平台下配置 VSL 链路的命令如下。

```
Switch 1(config)#vsl-aggregateport 1     // 进入 VSL 配置模式
Switch 1(config-vsu-ap)#port-member interface GigabitEthernet 0/1  copper
/* 将 Gi0/1 加入 VSL 组 1，如果是光口，则将 copper 更改为 fibber。需要注意的是，10.X 版本必须带介质，而 11.X 版本不需要 */
SW-1(config-vsu-ap)# port-member interface gigabitEthernet 0/2  copper
// 将 Gi0/2 加入 VSL 组 1，如果是光口，则将 copper 更改为 fibber
```

在 11.X 版本软件平台下配置 VSL 链路的命令如下。

```
Switch 1(config)# vsl-port 1
Switch 1(config-vsl-ap-1)# port-member interface TenGigabitEthernet 0/1
// 将 TE0/1 加入 VSL 组 1，需要注意是，10.X 版本必须带介质，而 11.X 版本不需要
Switch 1(config-vsl-ap-1)# port-member interface TenGigabitEthernet 0/2
// 将 TE0/2 加入 VSL 组 1
```

② 在组建 VSU 时，配置 BFD 检测的命令不同。

在 10.X 版本软件平台下配置 BFD 检测的命令如下。

```
Ruijie (config)#switch virtual domain 1    // 进入 VSU 参数配置
Ruijie(config-vs-domain)# dual-active detection bfd
// 打开 BFD 开关，其默认是关闭的
Ruijie(config-vs-domain)# dual-active pair interface Gi1/4/2 interface
```

任务 25　排除 VSU 故障

```
gi2/4/2
```
　　// 配置一对 BFD 检测接口

在 11.X 版本软件平台下配置 BFD 检测的命令如下。

```
Ruijie(config)#switch virtual domain 1    // 进入 VSU 参数配置
Ruijie(config-vs-domain)# dual-active detection bfd
```
　　// 打开 BFD 开关，其默认是关闭的

```
Ruijie(config-vs-domain)# dual-active bfd interface Gi1/4/2
```
　　// 配置一对 BFD 检测接口

```
Ruijie(config-vs-domain)# dual-active bfd interface Gi2/4/2
```

（10）常见故障排查 1：检查 VSU 系统状态。

先进行 VSU 相关状态检查，查看是否为 VSU 系统状态异常导致的断网。

① 检查 VSU 的配置是否正确，命令如下。

```
Ruijie#show switch virtual config
```

② 确定 VSL 链路是否有异常，命令如下。

```
Ruijie#show sw virtual link
Ruijie#show sw virtual link port
```
　　// 查看是否存在 OK 的 VSL 链路，如果存在，则说明 VSU 系统未分裂
```
Ruijie#show int count error
Ruijie#show int trans              //如果使用光口组建 VSL 链路，则查看光模块信息
Ruijie#show int trans dia          //如果使用光口组建 VSL 链路，则查看光纤链路光衰
```

③ 如果 VSL 链路全部下线，则分别登录两台设备，查看是否有一台设备已经进入恢复模式，使用如下命令。

```
Ruijie#show sw virtual dual-active summary
BFD dual-active detection enabled: Yes
Aggregateport dual-active detection enabled: NO
In dual-active recovery mode: Yes
```
　　/* 需要有一台设备进入恢复模式，如果两台设备都未进入恢复模式，那么原因就是 VSL 链路异常，并且没有配置双主机检测，VSU 系统出现了双主机 */

④ 如果有一台设备已经进入恢复模式，则需要检查拓扑，查看断网区域是否只在上连到进入恢复模式的主机箱上。

⑤ 如果 VSL 链路全部下线，则需要排查链路全部下线原因，主要操作如下。

　　a. 进入管理板与对应的 VSL 线卡，使用 "show exception" 命令查看出现故障时是否有死机信息。

　　b. 查看 LOG 是否有线卡 keepalive timeout 信息。

　　c. 查看 LOG 是否有管理板异常复位（非死机）情况发生。

（11）常见故障排查 2：检查发送到控制层面的 VSU 系统报文是否正常。

首先，通过使用 "show arp" 命令，查看不通的网段是否能够学习到 ARP。

其次，进行 CPP 查询，查看相关报文是否正常送入 CPU（网络中开启的对应协议，如连通性的协议必须查看，如 ARP、VRRP、BPDU 等）。

如下所示,每个命令连续收集5次。

```
Ruijie#show cpu-pro
Ruijie#show cpu-pro mb
Ruijie#show cpu-protect slot X
 /* 这里的X代表具体的槽位,如单机8610的4号槽,命令为"show cpu-pro slot 4";如
VSU中2号机箱4号槽位,命令为show cpu-pro slot 2/4 */
```

如果送线卡CPU或者送管理板CPU相关报文都一直显示为0,则说明对应报文送CPU时出现了异常,这可能是导致问题的原因。此外,10.4(3b17)之前版本的核心设备可能出现了NFPP时钟翻转问题,导致协议报文送管理板CPU异常。

(12)查看VSU系统的接口流量统计。

使用以下命令,可查看VSU系统的接口流量统计信息。

```
Ruijie#show int count rate    // 查看接口速率是否超过了接口带宽
Ruijie#show int count summary
 // 如果接口输入/输出广播报文极速增长,则基本上可以判断为网络中出现了环路
```

(13)检查VSU系统资源是否正常。

使用以下命令,可查看VSU系统资源是否正常。

```
Ruijie#show cpu         //查看VSU系统CPU资源是否正常
Ruijie#show memory      //查看VSU系统内存使用率
Ruijie#debug support
Ruijie(support)#show queue
//查看VSU系统队列情况,如果可用队列为0,则说明故障可能是因为相应丢列被耗尽而导致的
Ruijie(support)#show skb
 /* 查看VSU系统SKB队列,如果SKB队列可用值接近于0,则说明故障可能是因为SKB收发帧
队列被耗光而导致的  */
```

任务 26 使用 NFPP 技术保护交换机 CPU 不受攻击

【任务描述】

某企业网络的核心设备上连接有多台接入设备,以将企业网络中的 3000 多个用户设备接入到企业网中。其中,允许在核心交换机的一个接口上最多扩展出 200 台用户端设备。

为了避免网络中的终端设备感染病毒,防范来自内部网络的非法攻击,所有接入设备都需要启用"DHCP Snooping+DAI"安全检查,防范 ARP 欺骗。这些安全功能的启用会消耗接入交换机的 CPU 资源,需要调整 NFPP 参数,实现防网络攻击,优化网络传输效率。

【任务目标】

配置 NFPP 技术,保护交换机的 CPU 不受攻击。

【组网拓扑】

图 26-1 所示网络拓扑为某企业网络组网场景,由于核心网络中的汇聚设备下连接有 3000 台用户端设备,为防止遭受网络攻击,在网络中开启二层接入交换机端口安全功能,配置 NFPP 技术,保护核心网络中的三层交换机的 CPU 不受攻击。

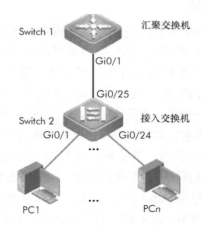

图 26-1 某企业网络组网场景

(备注:具体的设备连接接口信息可根据实际情况决定)

其中,在实施 NFPP 技术时,主要配置的内容包括以下几点。

(1)根据需求调整接入交换机 NFPP 功能参数,由于接入交换机上开启了 DAI 功能,网上发送的 ARP 报文接入交换机后,都需要送到 CPU 中进行处理。为了防止正常的 ARP 等报文被 NFPP 丢弃,需要关闭上连接口的 NFPP 相关功能,并调整接入交换机的 CPP 限速（默认限速180PPS，这在DAI动态监测的场景中偏小，需要增大参数）。

（2）根据核心网络中的设备连接的用户数量调整核心网络中的汇聚交换机上的NFPP属性，配置基于端口的限速/攻击检测参数。

（3）为了避免NFPP产生的日志太多，通过命令调整日志打印速率。

【设备清单】

二层交换机（1台），三层交换机（1台），网线（若干），PC（若干）。

【关键技术】

1. 什么是NFPP技术

在网络环境中经常会遇到网中病毒，以及一些恶意的攻击，这些攻击会给交换机带来过重的负担，引起交换机CPU利用率过高，导致交换机无法正常运行。例如，拒绝服务攻击，可能会大量消耗交换机内存、注册表项或者其他资源，使系统无法继续服务；又如，大量的报文流，涌向交换机的CPU，占用了全部CPU报文处理的带宽，导致正常的协议流和管理信息流无法被CPU处理，造成CPU震荡，从而导致数据的转发受影响，并使整个网络无法正常运行。

此时，可以通过网络基础保护策略（Network Foundation Protection Policy，NFPP）技术实施安全保障。NFPP技术可以有效地使系统免受这些攻击的影响。即使在系统受到攻击的情况下，也可保护系统各种服务正常运行，保持较低的CPU负载，从而保障整个网络的稳定运行。但作为保护交换机自身CPU不受攻击的手段，NFPP技术不能替代防范ARP欺骗等安全功能。

2. NFPP技术调整原则

设备默认支持并开启NFPP功能，建议保留并适当调整参数。

（1）在接入设备上通常无须调整，因为非网关设备没有配置网关IP地址，也不会运行一些复杂路由、管理协议等，无须额外消耗CPU，所以很少会遭受攻击。

（2）在汇聚设备上，默认实施基于接口的限速，在遭受100PPS/次攻击的情况下，200PPS阈值偏小。当接口下接入用户较多，遭受ARP攻击较多时，可能导致用户的正常ARP报文丢弃，导致用户端设备严重丢包，需要调整限速范围。在每个接口限速500PPS/次攻击的情况下，将其调整为800PPS较为合理，其他基于IP/MAC的限速攻击，检测阈值无须调整。

（3）隔离功能通常不建议开启。在汇聚设备上，如果攻击非常频繁，导致设备上的CPU资源消耗高达80%~90%，甚至90%以上，此时，可以考虑进行硬件隔离，并增大当前攻击的检测阈值，防止误判。因为一旦隔离，将会导致这些攻击性质的用户无法上网。

3. NFPP技术保障的应用

在日常网络基础保护策略应用中，NFPP主要保障的应用如下。

（1）ARP-Guard

汇聚交换机作为网关，需要正常处理用户端的ARP报文，ARP-Guard功能的主要目标为保护设备的CPU，防止大量ARP攻击报文发送到CPU，导致CPU的利用率升高，所以使用ARP-Guard实现了对发送到CPU上的ARP报文的限速和攻击检测。

任务 26 使用 NFPP 技术保护交换机 CPU 不受攻击

ARP-Guard 分为基于主机和基于物理接口两种防护方式。其中，基于主机防护又细分为基于源 IP 地址、基于 VLAN ID、基于物理接口和基于链路层源 MAC 地址等方式。每种攻击识别都有限速警戒线。例如，设置攻击告警水线为 8PPS。当 ARP 报文速率超过限速水线时，超限报文将被丢弃；当 ARP 报文速率超过攻击告警水线时，将输出警告信息。基于主机的攻击识别还可以对攻击源头采取硬件隔离措施。

（2）ARP 扫描

ARP 扫描是指链路层上的源 MAC 地址固定，而源 IP 地址变化；或者链路层上的源 MAC 地址和源 IP 地址固定，而目的 IP 地址不断变化，在其中实施 ARP 扫描。由于存在误判的可能，所以对检测出有 ARP 扫描嫌疑的主机不进行隔离，只是提供给网络管理人员参考。

（3）IP-Guard

当主机发出的报文目的地址为交换机直连网段上不存在或未上线用户的 IP 地址时，交换机会发出 ARP 请求。在某些特殊场景中，由于终端设备上有 ARP 攻击病毒，发送出过多的 ARP 广播请求，会导致设备 CPU 利用率升高。

IP-Guard 可以识别此类 ARP 攻击事件，并进行限速。其中，攻击识别也分为基于主机和基于物理接口两个类别。而基于主机通常采用源 IP 地址、VLAN ID、物理接口三者结合的方式识别。每种攻击识别都有限速水线和告警水线，当 IP 报文速率超过限速水线时，超限报文将被丢弃；当 IP 报文速率超过告警水线时，将输出警告信息，发送 TRAP 通知。同样，基于主机的攻击识别还会对攻击源头采取硬件隔离措施。

（4）NFPP 日志信息打印调整

当 NFPP 检测到攻击后，会在专用日志缓冲区中生成一条日志，以一定速率从专用缓冲区取出日志，生成系统消息，并从专用日志缓冲区中清除这条日志。

【实施步骤】

（1）在接入交换机上，配置交换机的防 ARP 欺骗功能。

① 创建 VLAN 10，划分用户到其中。

```
Ruijie#configure terminal
Ruijie(config)#vlan 10
Ruijie(config-vlan)#exit
Ruijie(config)#interface range GigabitEthernet 0/1-24
Ruijie(config-if-range)#switchport access vlan 10
Ruijie(config-if-range)#exit
Ruijie(config)#
```

② 在用户 VLAN 10 中配置 DHCP 安全信任端口。

```
Ruijie(config)#ip arp inspection vlan 10
Ruijie(config)#ip dhcp snooping
```

③ 在连接核心网络中三层汇聚交换机上连接口实施干道技术，配置防 ARP 安全、DHCP 安全。

```
Ruijie(config)#interface gigabitEthernet 0/25
Ruijie(config-if-GigabitEthernet 0/25)#switchport mode trunk
```

多层交换技术（实践篇）

```
Ruijie(config-if-GigabitEthernet 0/25)#ip dhcp snooping trust
Ruijie(config-if-GigabitEthernet 0/25)#ip arp inspection trust
Ruijie(config-if-GigabitEthernet 0/25)#exit
```

（2）在接入交换机上，配置交换机的 NFPP 功能。

① 在全局模式下，配置 NFPP。

默认情况下，交换机上的 NFPP 功能是开启的。二层交换机参数无须调整，只需将上连接口上的 NFPP 功能关闭即可。同时，由于开启了 DAI 的原因，需要调大 CPP 的参数值。

在全局模式下，使用如下命令配置 NFPP 功能。

```
Ruijie(config)#
Ruijie(config)#cpu-protect type arp pps 500
```

如果交换机上没有使用 DAI，则 CPP 参数无须调整。在全局模式下，调整 NFPP 参数如下。

```
Ruijie(config)#
Ruijie(config)#nfpp
Ruijie(config-nfpp)#log-buffer entries 1024
```
// 设置 NFPP LOG 缓存的容量为 1024 条（默认为 256 条）

```
Ruijie(config-nfpp)#log-buffer logs 1 interval 300
```
// 为避免 NFPP 产生的 LOG 太多，调整每次输出一条相同 LOG 信息的阈值为 300s

```
Ruijie(config-nfpp)#exit
```

② 在接口模式下，配置 NFPP。

为了防止网络中三层网关发送的正常报文（如正常 ARP 请求或回应报文）被接入交换机，误认为是攻击而丢弃，避免下连用户因无法获取网关的 ARP 信息而无法上网，需要将上连接口上的 NFPP 功能关闭。

```
Ruijie(config)#
Ruijie(config)#interface GigabitEthernet 0/25
Ruijie(config-if-GigabitEthernet 0/25)#no nfpp arp-guard enable
```
// 关闭接口的 ARP-Guard 功能，关闭该功能后，该接口收到的数据报不进行 NFPP 检测

```
Ruijie(config-if-GigabitEthernet 0/25)#no nfpp dhcp-guard enable
```
// 关闭接口的 DHCP-Guard 功能，关闭该功能后，该接口收到的数据报不进行 NFPP 检测

```
Ruijie(config-if-GigabitEthernet 0/25)#no nfpp dhcpv6-guard enable
```
// 关闭接口的 DHCPV6-Guard 功能，关闭该功能后，该接口收到数据报不进行 NFPP 检测

```
Ruijie(config-if-GigabitEthernet 0/25)#no nfpp icmp-guard enable
```
// 关闭接口的 ICMP-Guard 功能，关闭该功能后，该接口收到的数据报不进行 NFPP 检测

```
Ruijie(config-if-GigabitEthernet 0/25)#no nfpp ip-guard enable
```
// 关闭接口的 IP-Guard 功能，关闭该功能后，该接口收到的数据报不进行 NFPP 检测

```
Ruijie(config-if-GigabitEthernet 0/25)#no nfpp nd-guard enable
```
// 关闭接口的 ND-Guard 功能，关闭该功能后，该接口收到的数据报不进行 NFPP 检测

```
Ruijie(config-if-GigabitEthernet 0/25)#exit
```

（3）在核心网络中汇聚交换机上，配置 NFPP 功能。

```
Ruijie(config)#
Ruijie(config)#nfpp
```

任务 26　使用 NFPP 技术保护交换机 CPU 不受攻击

```
Ruijie(config-nfpp)#arp-guard attack-threshold per-port 800
                  // 设置每个接口的攻击阈值为 800 个，超过此值时丢弃并输出攻击日志
Ruijie(config-nfpp)#arp-guard rate-limit per-port 500
                  // 每个接口每秒限速 500 个 ARP 报文，多余 ARP 报文将被丢弃
                  // 默认限速阈值是 100 个，在实际应用时该数值偏小
Ruijie(config-nfpp)#log-buffer entries 1024
                  // 设置 NFPP LOG 缓存的容量为 1024 条（默认为 256 条）
Ruijie(config-nfpp)#log-buffer logs 1 interval 300
//调整 LOG 300s 输出 1 次
Ruijie(config-nfpp)#exit
```

【注意事项】

　　如果需要开启硬件隔离功能，为了防止误判，应增大限速及攻击检测的阈值。需要注意的是，隔离功能通常不建议开启，因为一旦隔离将会导致这些攻击性质的用户无法上网。在汇聚层的交换机上，如果用户攻击非常频繁，导致设备上的 CPU 利用率达到 90% 以上，此时可以考虑进行硬件隔离，并增大当前攻击的检测阈值，防止误判。

```
Ruijie(config)#
Ruijie(config)#nfpp              // 进入 NFPP 配置模式
Ruijie(config-nfpp)#arp-guard isolate-period 600
                  // 超过 ARP 攻击阈值后，对用户进行隔离，设置隔离时间为 600s
Ruijie(config-nfpp)#arp-guard attack-threshold per-src-mac 30
   /* 设置 MAC 攻击阈值为 30 个，如果交换机检测单台设备发送的 ARP 报文大于 30 个，那么交换
机会把该用户放入 ARP 攻击表，可以对这些用户进行硬件隔离（默认不进行硬件隔离），可以配置隔离
时间，设置隔离时间后会占用硬件表项资源。默认攻击阈值是 8 个 */
Ruijie(config-nfpp)#arp-guard attack-threshold per-src-ip 30
   /* 设置每个 IP 的攻击阈值为 30 个，如果交换机检测到单台 IP 主机上发送的 ARP 报文大于 30 个，
那么交换机会把该用户放入 ARP 攻击表，可以对这些用户进行硬件隔离。默认每个 IP 的攻击阈值是 8 个 */
Ruijie(config-nfpp)#arp-guard rate-limit per-src-mac 20
   // 每个 MAC 每秒限速 20 个 ARP 报文，多余的 ARP 报文将被丢弃（默认限速阈值是 4 个）
Ruijie(config-nfpp)#arp-guard rate-limit per-src-ip 20
   // 每个 IP 每秒限速 20 个 ARP 报文，多余的 ARP 报文将被丢弃（默认限速阈值是 4 个）
Ruijie(config-nfpp)#ip-guard attack-threshold per-src-ip 80
                                            // 设置 IP 攻击阈值为 80 个
Ruijie(config-nfpp)#ip-guard isolate-period 600
                  // 超过 IP 攻击阈值后，对用户进行隔离，设置隔离时间为 600s
```

【测试验证】

　　（1）查看 ARP-Guard 的配置情况（不同版本软件的阈值可能有所区别，具体以实际查询结果为准），如图 26-2 所示。

　　（2）查看 ARP-Guard 的扫描表，如图 26-3 所示。

　　（3）查看 ARP-Guard 被隔离的用户，如图 26-4 所示。

```
Ruijie#show nfpp arp-guard summary
(Format of column Rate-limit and Attack-threshold is per-src-ip/per-src-mac/per-port.
Interface  Status   Isolate-period Rate-limit      Attack-threshold Scan-threshold
Global     Enable   600            6/6/200         10/10/400        2000
Gi0/25     Disable  -              -/-/-           -/-/-            -
Maximum count of monitored hosts: 1000
Monitor period: 600s
```

- Enable → 全局开启
- 600 → 隔离周期 10min

图 26-2　查看 ARP-Guard 的配置情况

```
Ruijie#show nfpp arp-guard scan
VLAN  interface  IP address  MAC address     timestamp
----  ---------  ----------  ------------    ---------
1     Gi0/6      -           00d0.f8aa.bb36  2013-3-16 9:23:43
1     Gi0/6      -           001a.a942.f2df  2013-3-16 9:24:41
```

图 26-3　查看 ARP-Guard 的扫描表

```
Ruijie#show nfpp arp-guard hosts
If col_filter 1 shows '*', it means "hardware do not isolate host".
VLAN  interface  IP address  MAC address     remain-time(s)
----  ---------  ----------  ------------    --------------
1     Gi0/6      -           001a.a942.f2df  133
1     Gi0/6      -           00d0.f8aa.bb36  38
Total: 2 hosts
```

- 133 / 38 → 该值指剩余隔离时间值，不是已经隔离时间值

图 26-4　查看 ARP-Guard 被隔离的用户

（4）查看放入日志缓冲区的用户，如图 26-5 所示。

```
Ruijie#show nfpp log buffer
Protocol VLAN Interface IP address   MAC address     Reason
-------- ---- --------- ----------   ------------    ------
ARP      1    Gi0/6     -            00d0.f8aa.bb36  DoS
ARP      1    Gi0/6     -            001a.a942.f2df  DoS
ARP      1    Gi0/6     -            00d0.f8aa.bb36  DoS
ARP      1    Gi0/6     -            001a.a942.f2df  DoS
ARP      1    Gi0/6     -            00d0.f8aa.bb36  DoS
ARP      1    Gi0/6     -            001a.a942.f2df  DoS
ARP      1    Gi0/6     -            001a.a942.f2df  SCAN
```

被检测出用户放入 log 缓冲区，通过该命令查看

图 26-5　查看放入日志缓冲区的用户

任务 ㉗ IP Source Guard 防范攻击

【任务描述】

某企业网络为方便地址管理，在网络中心的核心交换机上创建了 DHCP Server，用户的网关也配置在核心交换机上，接入交换机上连接的用户计算机使用 DHCP 动态获取 IP 地址。为了防止内网用户私设 IP 地址，拟实施 IP Source Guard 功能，禁止私设 IP 地址的用户访问外网。

IP Source Guard 功能可以防止用户私设 IP 地址及改变源 IP 地址的扫描行为，要求用户必须以 DHCP 方式动态获取 IP 地址，否则将无法使用网络。此外，通过 IP Source Guard 并配合 ARP-Check 功能，可以有效预防 ARP 欺骗。

【任务目标】

配置 IP Source Guard 防范攻击。

【组网拓扑】

图 27-1 所示网络拓扑为某企业网络 DHCP 配置场景，通过配置 IP Source Guard 功能防范 DHCP 攻击。

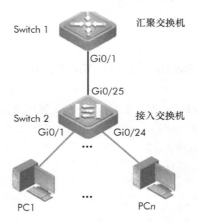

图 27-1 某企业网络 DHCP 配置场景
（备注：具体的设备连接接口信息可根据实际情况决定）

其中，为了防范 DHCP 攻击，需要配置的内容包括以下几点。

（1）在三层汇聚交换机上开启 DHCP Server 功能（部分场景中，客户可能采用专用 DHCP 服务器，此时，三层汇聚交换机只需要启用 DHCP Relay 功能即可）。

多层交换技术（实践篇）

（2）全局模式下，在二层接入交换机上开启 DHCP Snooping 功能，将二层接入交换机上连三层汇聚交换机的干道接口，配置为 DHCP Snooping 信任口。

（3）在二层接入交换机的连接用户的接口上，开启 IP Source Guard 功能。

（4）企业防护网中个别用户使用静态 IP 地址，配置 IP Source Guard 功能实现安全防护。

【设备清单】

二层交换机（1台），三层交换机（1台），网线（若干），PC（若干）。

【关键技术】

IP Source Guard（IP 源防护）通过维护一个 IP 源地址绑定数据库，可以在对应的接口上对收到的主机的报文进行基于源 IP 和源 MAC 的报文过滤，保证只有 IP 源地址绑定在数据库中的主机才能正常使用网络。

IP Source Guard 功能开启后，自动将 DHCP Snooping 安全绑定到数据库中的合法用户接入口上，并将 IP Source Guard 保护的 IP 源地址同步绑定到数据库（硬件安全表项）中。这样，IP Source Guard 就可以在打开 DHCP Snooping 的设备上，对收到的客户端上发出的报文进行严格过滤。

默认情况下，开启 IP Source Guard 功能的接口会过滤所有非 DHCP 方式获取的 IP 报文；只有当客户端通过 DHCP 技术从服务器获取到合法的 IP 地址，或者管理员为客户端配置了静态 IP 地址绑定时，接口才允许和这个客户端匹配的 IP 报文通过。

IP Source Guard 功能支持基于"IP+MAC"或者基于 IP 地址的过滤，如果打开基于"IP+MAC"的过滤，则 IP Source Guard 会对所有报文的"IP+MAC"进行检测，仅仅允许 IP 源地址绑定表中存在的用户的报文通过；而基于 IP 的过滤，仅仅会对报文的源 IP 地址进行检测。

【实施步骤】

（1）在三层汇聚交换机 Switch 1 上开启 DHCP 服务。

```
Ruijie(config)#
Ruijie(config)#interface VLAN 1          // 配置三层汇聚交换机管理 IP 地址
Ruijie(config-if-VLAN 1)#ip address 192.168.1.254 255.255.255.0
// 该 IP 地址也为用户默认网关地址
Ruijie(config-if-VLAN 1)#exit

Ruijie(config)#service dhcp                                // 开启 DHCP 服务

Ruijie(config)#ip dhcp pool vlan1-IP                       // 创建 DHCP 地址池
Ruijie(dhcp-config)#network 192.168.1.0 255.255.255.0      // 可分配的地址段
Ruijie(dhcp-config)#dns-server 218.85.157.99               // 分配 DNS 地址
Ruijie(dhcp-config)#default-router 192.168.1.254           // 分配给用户网关地址
Ruijie(dhcp-config)#exit
```

任务 ㉗ IP Source Guard 防范攻击

（2）在二层接入交换机 Switch 2 上开启 DHCP Snooping 功能。

```
Ruijie(config)#
Ruijie(config)#service dhcp                  // 开启 DHCP 服务
Ruijie(config)#ip dhcp snooping              // 开启 DHCP Snooping 功能
Ruijie(config)#interface GigabitEthernet 0/25  // 打开连接 DHCP 服务器的上连接口
Ruijie(config-GigabitEthernet 0/25)#ip dhcp snooping trust
//将连接 DHCP 服务器的接口配置为可信任口
Ruijie(config-GigabitEthernet 0/25)#exit
/* 开启 DHCP Snooping 功能，交换机所有接口默认为 untrust 口，交换机只转发从 trust
接口上收到的 DHCP 响应报文（如 OFFER、ACK 报文） */
```

（3）在二层接入交换机 Switch 2 连接用户的接口上开启 IP Source Guard 功能。

```
Ruijie(config)#
Ruijie(config)#interface range GigabitEthernet 0/1-2
Ruijie(config-if-range)#ip verify source port-security
Ruijie(config-if-range)#exit
/* 开启基于源 "IP+MAC" 数据报文检测安全功能，将 DHCP Snooping 形成的 snooping 映射表
写入到地址绑定数据库中，须正确配置 "ip verify source port-security"，不要使用 "ip verify
source"（仅绑定 IP）命令，因为部分产品存在限制，只绑定 IP 的情况下可能出现异常 */
/* 需要注意的是，如果交换机下还有级联交换机，则不要在级联接口配置 "IP Source Guard"，
而应该在接入交换机上部署 IP Source Guard 功能，避免两台设备都硬件绑定用户的 "IP+MAC"，消
耗设备硬件资源表项 */
```

（4）在二层接入交换机 Switch 2 上配置静态绑定用户安全。

在二层接入交换机上配置静态绑定用户，这些用户希望采用静态 IP 地址，并实现安全检查，避免接口下其他用户私设 IP 地址。

```
Ruijie(config)#
Ruijie(config)#ip source binding 001a.a2bc.3a4d vlan 10 192.168.10.5
interface GigabitEthernet 0/15
                 // 在 Gi0/15 接口上绑定 MAC 地址、IP 地址
Ruijie(config)#interface GigabitEthernet 0/15  // 打开连接用户的某一台计算机的接口
Ruijie(config-if-GigabitEthernet t 0/15)#ip verify source port-security
                 // 开启基于源 "IP+MAC" 的报文检测安全防护功能
Ruijie(config-if-GigabitEthernet t 0/15)#exit
```

【测试验证】

（1）查看三层汇聚交换机 DHCP 服务器地址池分配情况，如图 27-2 所示。

图 27-2 查看三层汇聚交换机 DHCP 服务器地址池分配情况

（2）在接入计算机上，查看测试计算机获取 IP 地址的情况；在 DOS 命令行中，使用 "ipconfig/all" 命令，查看活动的 IP 地址信息，如图 27-3 所示。

图 27-3　查看测试计算机获取 IP 地址的情况

（3）查看 DHCP Snooping 表，如图 27-4 所示。

图 27-4　查看 DHCP Snooping 表

（4）查看 IP Source Guard 相关信息，如图 27-5 所示。

图 27-5　查看 IP Source Guard 相关信息

（5）测试网络连通性，验证自动获取 IP 地址（获取的 IP 地址是 192.168.1.1）的计算机能否 Ping 通核心交换机上的网关地址，如图 27-6 所示。

图 27-6　测试网络连通性

（6）查看网关的 MAC 地址，如图 27-7 所示。

图 27-7　查看网关的 MAC 地址

任务 28 保护交换机端口安全

【任务描述】

某企业的网络中心为了实施网络安全策略,禁止非授权的用户接入访问公司内网,希望接入交换机上接入用户的 IP 地址和 MAC 地址必须是管理员指定的,或者希望使用者能够在固定端口上网,以减少网络攻击事件发生,防护网络安全。

【任务目标】

保护交换机端口安全。

【组网拓扑】

图 28-1 所示网络拓扑为某企业网络的用户接入网络场景,需要实施交换机的端口安全防护,防范网络攻击事件发生。

图 28-1 某企业网络的用户接入网络场景
(备注:具体的设备连接接口信息可根据实际情况决定)

其中,在 Fa0/1 及 Fa0/24 接口上开启交换机端口安全功能,并且限制 MAC 地址绑定的条目为 1。要求 PC1(IP 地址为 192.168.1.1/24,MAC 地址为 0021.CCCF.6F70)只能接在交换机的 Fa0/1 接口上,并且做 "IP+MAC" 地址绑定,其他计算机接入该接口将无法通信;要求 Fa0/2 接口只能接入 IP 地址为 192.168.1.2/24、MAC 地址不受限制的计算机,其他 IP 地址的计算机都不能从该接口接入。

多层交换技术（实践篇）

【设备清单】

二层交换机（1台），三层交换机（1台），网线（若干），PC（若干）。

【关键技术】

交换机上的端口安全功能通过定义报文的源 MAC 地址来限定收到的报文是否可以进入交换机的接口，可以通过静态方式设置特定的 MAC 地址，或者限定动态学习到的 MAC 地址的个数，来控制报文是否可以进入交换机，其中，实施了端口安全功能的端口称为安全端口。

配置了端口安全功能的交换机只允许源 MAC 地址为端口安全地址表中配置或者学习到的 MAC 地址的报文进入，其他报文将被丢弃。网络管理人员还可以设定端口安全地址，如绑定 "IP+MAC" 地址，或者仅绑定 IP 地址，限制只有安全地址为源 MAC 地址的报文才能进入交换机。

通过实施端口安全功能，可以控制端口下接入用户的 IP 和 MAC 必须是管理员指定的合法地址，或者控制用户在固定端口下上网，或者控制端口下的用户 MAC 数，以防止 MAC 地址耗尽安全攻击事件发生。例如，MAC 地址攻击病毒通过发送持续变化的构造出来的 MAC 地址，交换机短时间内学习大量无用的 MAC 地址，一台交换机上 8KB/16KB 的 MAC 地址表会很快填充满，并无法学习合法用户的 MAC 地址，导致交换机通信异常。

【实施步骤】

（1）在二层接入交换机上配置 Fa0/1 接口安全。

按照安全配置要求，规定 Fa0/1 接口只能接入 IP 地址是 192.168.1.1/24，且 MAC 地址是 0021.CCCF.6F70 的计算机。

```
Ruijie#configure terminal
Ruijie(config)#interface FastEthernet 0/1
Ruijie(config-if-FastEthernet 0/1)#switchport port-security
// 开启端口安全功能
Ruijie(config-if-FastEthernet0/1)#switchport port-security binding 0021.CCCF.6F70 vlan 1 192.168.1.1
/* 把属于 VLAN 1，且 MAC 地址为 0021.CCCF.6F70、IP 地址为 192.168.1.1/24 的计算机绑定在此接口上 */
Ruijie(config-if-FastEthernet 0/1)#exit
```

（2）在接入交换机上配置 Fa0/2 接口安全地址。

按照安全配置要求，规定 Fa0/2 接口只能接入 IP 地址是 192.168.1.2/24，且 MAC 地址无要求的计算机，即 Fa0/2 下只能连接 IP 地址为 192.168.1.2/24 的计算机。

```
Ruijie(config)#
Ruijie(config)#interface FastEthernet 0/2
Ruijie(config-if-FastEthernet 0/2)#switchport port-security
// 开启端口安全功能
Ruijie(config-if-FastEthernet 0/2)# switchport port-security binding
```

任务 ㉘ 保护交换机端口安全

```
192.168.1.2
                        // 把 IP 地址是 192.168.1.2 的终端绑定在交换机的第二个接口上
Ruijie(config-if-FastEthernet 0/2)#exit
```

【测试验证】

查看二层接入交换机绑定安全表项,如图 28-2 所示。

图 28-2 查看二层接入交换机绑定安全表项

【注意事项】

二层交换机上的端口安全功能分为"IP+MAC"绑定、仅 IP 绑定、仅 MAC 绑定三种类型,通过如下命令进行设定。

```
Ruijie(config-if-FastEthernet 0/1)#switchport port-security binding  ?
A.B.C.D  IP address
H.H.H    48-bit hardware address
X:X:X:X::X  IPv6 address
```

如果需要绑定"IP+MAC",则可以使用如下命令。

```
Ruijie(config-if-FastEthernet 0/1)#switchport port-security binding 0021.
CCCF.6F70 vlan 1 192.168.1.1
```

如果需要仅绑定 IP,则可以使用如下命令。

```
Ruijie(config-if-FastEthernet 0/1)#switchport port-security binding 192.
168.1.2
```

如果交换机仅做 MAC 绑定,则可以使用如下命令。

```
Ruijie(config-if-FastEthernet 0/1)#switchport port-security mac-address
0021.CCCF.6F70
```

此外,在交换机上设置"IP+MAC"绑定或仅 IP 绑定后,交换机还是会动态学习到接口下连接的用户设备 MAC 地址。例如,在交换机端口上做如下绑定。

```
Ruijie(config-if-FastEthernet 0/1)#switchport port-security binding 1414.
4b19.ecc1 vlan 1 192.168.1.1
Ruijie(config-if-FastEthernet 0/1)#switchport port-security
```

此时,查看到的配置信息如下。

```
Ruijie#show port-security address
Vlan Mac Address IP Address Type Port Remaining Age (mins)
---- ----------- -------------------------- ---------- ------ ----
1 1414.4b19.ecc1 192.168.1.1 Configured Fa0/1 -
```

可见,当用户接入后,交换机依然会动态学习用户的 MAC 地址,可使用以下命令进行查看。

多层交换技术（实践篇）

```
Ruijie#show port-security address
Vlan Mac Address IP Address Type Port Remaining Age (mins)
---- ---------- -------------------- ---------- -------- --------------
1 1414.4b19.ecc1 Dynamic Fa0/1 -
1 1414.4b19.ecc1 192.168.1.1 Configured Fa0/1 -
```

如果要让"IP+MAC"绑定或仅 IP 绑定的用户生效，则必须先让端口安全学习到用户的 MAC 地址。例如，交换机做如下设置，那么 IP 地址为 192.168.1.2 的用户无法上网。

```
interface fastEthernet 0/1
switchport port-security mac-address 1414.4b19.0000 vlan 1
switchport port-security binding 1414.4b19.ecc1 vlan 1 192.168.1.2
switchport port-security maximum 1
switchport port-security
```

其原因是 MAC 地址最大允许数是 1，已经绑定了一个 MAC 地址为 1414.4b19.0000 的设备，所以端口安全不能再学习到 MAC 地址了。而端口安全中的"IP+MAC"（或 IP）绑定生效时必须要先学习到端口上配置的安全 MAC 地址，所以 IP 地址为 192.168.1.2/24 的用户无法通过交换机上安全策略。如果要让该用户上网，则可以通过如下命令绑定该用户的 MAC 地址来实现。

```
interface fastEthernet 0/1
switchport port-security mac-address 1414.4b19.ecc1 vlan 1
switchport port-security mac-address 1414.4b19.0000 vlan 1
switchport port-security binding 1414.4b19.ecc1 vlan 1 192.168.1.2
switchport port-security maximum 2
switchport port-security
```

任务 29 多对一镜像保护网络安全

【任务描述】

某学院最近发现在上课期间，教学楼网络经常发生堵塞现象，使得整个楼层所有多媒体教室的网络非常缓慢。网络中心管理员决定利用端口镜像（SPAN）技术，来逐间排查多媒体教室，定位异常数据流量的来源。

端口镜像技术能将指定端口上收到的报文复制到交换机上另一个连接网络监测设备的端口上，进而实现网络监控与流量分析。

【任务目标】

配置多对一镜像安全。

【组网拓扑】

图 29-1 所示网络拓扑为某学院多媒体教室的网络接入场景，通过配置网络中的三层汇聚交换机上的连接的端口镜像监控服务器，监控 Gi0/1 及 Gi0/2 接口上入方向和出方向的数据流，逐间教室排查网络异常。

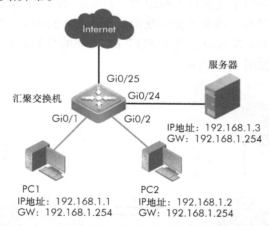

图 29-1 某学院多媒体教室的网络接入场景
（备注：具体的设备连接接口信息可根据实际情况决定）

【设备清单】

三层交换机（1台），网线（若干），PC（若干）。

【关键技术】

端口镜像是利用交换机的接口具有的 SPAN 镜像技术提供的功能，将指定接口上收到

多层交换技术（实践篇）

的报文复制到交换机上另一个连接有网络监测功能的接口上，进行网络监控与流量分析。通过 SPAN 镜像技术可以监控所有从源接口进入和输出的报文，实现报文快速、原封不动地"复制"。

在实施端口镜像技术的过程中，交换机不会改变镜像报文的任何信息，也不会影响原有报文的正常转发。同时，对于源/目的接口的介质类型没有要求，可以是光口镜像到电口，也可以是电口的流量镜像到光口；对于源/目的接口的属性没有要求，可以是 Access 端口镜像到 Trunk 口，也可以是 Trunk 口镜像到 Access 端口。

当需要把一台交换机的一个口或者多个接口上的流量，镜像（复制）到本交换机上某个接口的时候，就可以考虑采用多对一镜像技术。

需要注意的是，端口镜像中的源接口可以是多个接口的双向流量一起镜像到目的接口，也可以是部分接口上的入方向（RX）流量与部分接口上的出方向（TX）流量结合起来镜像到某个目的接口。但是目的接口只能有一个，即监控的服务器或者网络设备只能有一台。

端口镜像的模式有很多种，其中，多对一镜像技术通常用于网络中安装有监控服务器（如数据库操作审计服务器、日志记录服务器、上网行为管理服务器、流量统计或者监控服务器）、视频加速缓存设备（如 Powercache）等设备的情况。故障定位时，若要采用抓包分析（如目的计算机上安装有 Wireshark、Sniffer 软件），则需要采用端口镜像功能。

虽然多对一镜像的目的接口只能有一个，即目的服务器只能接一台，但是有一种特别的运用场景需要注意——上网行为管理设备以旁路的方式接到交换机设备的场景，该设备做透明模式工作，需要一个接口收数据，一个接口发数据，监控网络上下行流量的模型，从而分析每个用户上网的流量行为。所以，该设备需要两个接口接到一台交换机上（标记为 A、B），此时需要交换机上连接口（标记为 C）入方向的数据镜像到 A 接口，出方向的数据镜像到 B 接口，需要在交换机上面创建两个 Monitor Session，Session 1 将 C 接口 RX 方向的数据镜像到 A 接口；Session 2 将 C 接口 TX 方向的数据镜像到 B 接口，从而实现客户需求，参考组网拓扑如图 29-2 所示。

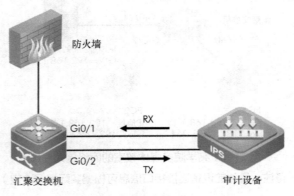

图 29-2　参考组网拓扑

【实施步骤】

```
Ruijie#configure terminal
Ruijie(config)#monitor session 1 source interface GigabitEthernet 0/1 both
```

任务 29 多对一镜像保护网络安全

```
/* 指定端口镜像上的源接口为 Gi0/1，即被监控接口，交换机可以指定多个源接口，这里的 both
表示双方向的数据流 */
    Ruijie(config)#monitor session 1 source interface GigabitEthernet 0/1 rx
    // 在实施过程中，如果只需要镜像交换机入方向的数据流，则应将 both 关键字改为 rx 关键字
    Ruijie(config)#monitor session 1 source interface GigabitEthernet 0/2 tx
    // 如果只需要镜像交换机出方向的流量，则可将 both 关键字改为 tx 关键字
    Ruijie(config)#monitor session 1 destination interface GigabitEthernet 0/24 switch
    // 指定 Gi0/24 口为端口镜像目的接口，即监控接口。关键字 switch 表示目的接口也能够上网
```

【测试验证】

（1）使用"show monitor"命令，查看端口镜像的状态。

```
Ruijie#show monitor
sess-num:1                                   // session 号为 1
span-type:LOCAL_SPAN                         // 配置的是本交换机端口镜像
src-intf:
GigabitEthernet 1/2   frame-type Both        // 镜像 Gi0/2 的输入和输出方向数据流
src-intf:
GigabitEthernet 1/1   frame-type Both        // 镜像 Gi0/1 的输入和输出方向数据流
dest-intf:
GigabitEthernet 1/24                         // 镜像的目的端口
mtp_switch on                                // 镜像的目的端口能转发数据流
```

（2）测试网络连通性，如图 29-3 所示，确认监控服务器能否访问外网。

图 29-3 测试网络连通性

备注：日常网络应用中，经常混淆使用"接口"和"端口"两个术语，在没有严格区分的情况下，广义上默认其可以交叉使用。但两者狭义上还有细微区别。其中，接口来自"interface"，多指物理口，端口来自"port"，多指逻辑口。

任务 ㉚ 配置交换网络中的一对多镜像

【任务描述】

某公司的网络最近异常缓慢,网络中心决定利用端口镜像功能,将指定接口的报文复制到交换机上另一个连接有网络监测设备的接口,监控所有从源接口进入和输出的报文。一对多(多对多)的 SPAN 技术还可以将一个接口的流量镜像到多个目的接口。

【任务目标】

配置交换网络中的一对多镜像。

【组网拓扑】

图 30-1 所示网络拓扑为某企业网络工作场景。其中,Gi0/1 及 Gi0/2 接口连接用户,Gi0/21 及 Gi0/22 接口连接两台监控服务器,监控服务器 1 及监控服务器 2 都能监控到 Gi0/1 及 Gi0/2 接口上连接设备的数据流。

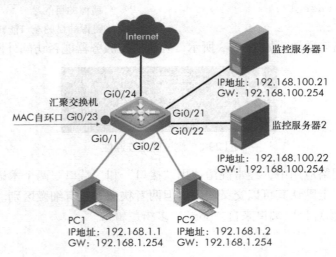

图 30-1 某企业网络工作场景
(备注:具体的设备连接接口信息可根据实际情况决定)

【设备清单】

三层交换机(1 台),网线(若干),PC(若干)。

【关键技术】

在实施一对多端口镜像的过程中,需要注意的是,一对多可以将一个接口的流量镜像

任务 ㉚ 配置交换网络中的一对多镜像

到多个目的接口，目前只有锐捷网络的 N18K、S86E、S78E 等系列交换机才支持一对多（多对多）的端口镜像。对于不支持一对多的设备，可以通过普通镜像功能（不使用自环口）将 RSPAN 目的接口连接到一台没有实施安全配置的交换机上，再划分多个接口到一个 VLAN（RSPAN 使用的 VLAN）中，并关闭所有接口的 MAC 地址学习功能（no mac-address-learning）和接口风暴控制（类似 Hub 泛洪，由于没有进行 MAC 地址学习，报文才会进行泛洪，所以需要关闭风暴控制防止镜像流量太多被抑制而丢弃），进而实现一对多功能。

在实施端口镜像的过程中，需要注意的事项包括以下几点。

（1）在核心网络中的三层汇聚交换机上创建 Remote VLAN。

（2）指定核心网络中的三层汇聚交换机为 RSPAN 的源设备，配置直连服务器的接口 Gi0/21 及 Gi0/22 为镜像源接口；选择一个 DOWN 状态的端口（本例为 Gi0/23）为镜像输出接口，将该接口加入 Remote VLAN，并配置为 MAC 自环。

（3）将直连 PC1 和 PC2 的接口加入 Remote VLAN。

此外，还需要重点说明的是，在网络中的三层汇聚交换机上需要将一个未使用的接口配置为 MAC 自环口，配置为 MAC 自环口后，该接口不插网线或光线，接口会自动 UP，并且接口状态灯亮，显示绿色。Lookback 口不能做其他配置，否则可能会导致监控服务器无法接收到监控数据流。

【实施步骤】

（1）在核心网络中的三层汇聚交换机上配置 Remote VLAN，这个 VLAN 需要交换机上没有使用的业务 VLAN。

```
Ruijie#configure terminal
Ruijie(config)#vlan 100              // 设置一个 VLAN 承载监控流量
Ruijie(config-vlan)#remote-span      // 设置一个 SPAN VLAN
Ruijie(config-vlan)#exit
```

（2）在核心网络中的三层汇聚交换机上配置 RSPAN 源设备。

```
Ruijie(config)#
Ruijie(config)#monitor session 1 remote-source
                             // 创建 RSPAN Session 1，指定该设备为源设备
Ruijie(config)#monitor session 1 source interface GigabitEthernet 0/1 both
                             // 配置 Gi0/1 为源接口，镜像双向数据流
Ruijie(config)#monitor session 1 source interface GigabitEthernet 0/2 both
                             // 配置 Gi0/2 为源接口，镜像双向数据流
```

（3）在核心网络中的三层汇聚交换机上指定自环口 Gi0/23 为镜像的目的接口。

```
Ruijie(config)#
Ruijie(config)#monitor session 1 destination remote vlan 100 interface GigabitEthernet 0/23 switch
    // 在交换机上指定 Gi0/23 为镜像输出接口，并加入 Remote VLAN 100
    /* 如果将流量引入 Loopback 接口，会强制携带 switch 关键字。如果目的口没有与外界通信的需求，则推荐不配置该 switch 参数 */
Ruijie(config)#
```

多层交换技术（实践篇）

```
Ruijie(config)#interface GigabitEthernet 0/23
Ruijie(config-if-GigabitEthernet 0/23)#switchport access vlan 100
Ruijie(config-if-GigabitEthernet 0/23)#mac-loopback
                    // 配置MAC自环口，可以看到原先DOWN的接口直接变为UP
                    // 自环口不要再做其他配置，也不要连接线缆

Switch A(config-if-GigabitEthernet 0/23)#no mac-address-learning
            // 务必增加该命令，否则端口镜像只能到自环口，不能镜像到目的接口
            // 使用"show int counters rate"命令，确认接口数据流量是否镜像到目的接口
Ruijie(config-if-GigabitEthernet 4/23)#end

Ruijie#clear mac-address-table dynamic interface GigabitEthernet 0/23
                    // 配置完成后需要清除自环口的MAC地址表
```

（4）在核心网络中的三层汇聚交换机上，将监控服务器的接口 Gi0/21 及 Gi0/22 加入 Remote VLAN。

```
Ruijie(config)#
Ruijie(config)#interface range GigabitEthernet 0/21-22
Ruijie(config-if-range)#switchport access vlan 100
                    // 配置交换机的接口Gi0/21和Gi0/22属于Remote VLAN 100
Ruijie(config-if-range)#exit
```

备注1：如果设备上开启了生成树协议，同时有其他 Trunk 口，由于 RSPAN 的镜像中的目的接口有 MAC-Loopback 功能，会导致流量在 Remote VLAN 中打环，所以需要在所有 Trunk 口上做 VLAN 修剪，如本例中的 remove vlan 100。

备注2：在指定自环口 Gi0/23 为镜像目的接口的配置过程中，如果配置了 switch 关键字，那么需要在自环口关闭 MAC 地址学习功能，并清除自环口的 MAC 地址表项；如果没有配置 switch 关键字，则不需要关闭自环口的 MAC 地址学习功能。

【测试验证】

查看端口镜像的状态，如图30-2所示。

图30-2 查看端口镜像的状态

任务 31 配置交换机基于流量的镜像安全

【任务描述】

某公司的网络最近异常缓慢，但在进行网络故障排查时，由于接口上的流量太大，如果对普通接口实施镜像安全，将导致计算机性能难以分析，很难捕捉到希望抓取的特定流量报文（如某个 MAC 的流量，或者某个源 IP 访问某个目的 IP 的流量）；或者接口上的流量太大，网络中部署的监控服务器、日志审计服务器等无法分析所有数据，只希望抓取特定的流量报文。此时都可以使用基于流量的镜像功能。

【任务目标】

配置交换机基于流量的镜像。

【组网拓扑】

图 31-1 所示网络拓扑为某企业网场景，实施基于流量的镜像安全防范。

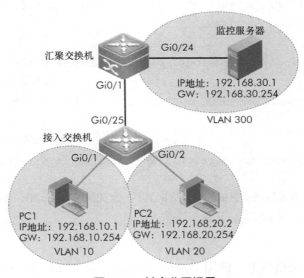

图 31-1 某企业网场景

（备注：具体的设备连接接口信息可根据实际情况决定。）

实施基于流量的镜像安全防范，需要在三层汇聚交换机上实施如下配置。

（1）在三层汇聚交换机上配置 ACL，允许用户网段 192.168.10.0/24 访问。

（2）在三层汇聚交换机上配置接口镜像功能，将二层接入交换机的接口 Gi0/1 设置为

多层交换技术（实践篇）

接口镜像的源接口，并关联 ACL。

（3）将连接监控服务器的接口 Gi0/24 设置为接口镜像的目的接口。

【设备清单】

二层交换机（1 台），三层交换机（1 台），网线（若干），PC（若干）。

【关键技术】

基于流量的镜像接口安全，通过 ACL 定义感兴趣的流量（如某个特定网段的 IP 报文、HTTP 网页流量等），实施流量监控。

流量监控技术依托交换机上的 ACL 技术实现，三层交换设备上的 ACL 功能丰富，支持基于二层帧类型、MAC 地址、IP 地址、TCP/UDP 端口，甚至 ACL80（报文的前 80 个字节）等来匹配流量，再由定义好的 ACL 关联镜像，捕捉到源接口上 ACL 所关注的流量，最后镜像到目的接口，而其他流量不会被镜像。

需要注意的是，目前交换机支持的基于流量的镜像只能实现 RX 方向（接口入方向），无法实现 TX（接口出方向）及 Both 方向的监控。

【实施步骤】

在三层汇聚交换机上完成基于流量的镜像配置。

```
Ruijie#configure terminal
Ruijie(config)#ip access-list extended ruijie      // 创建 IP ACL，名称为 ruijie
Ruijie(config-ext-nacl)#permit ip 192.168.10.0 0.0.0.255 any
Ruijie(config-ext-nacl)#exit

Ruijie(config)#monitor session 1 source interface GigabitEthernet 0/1 tx
                                                   // 监控从 Gi0/1 口出去的所有流量
Ruijie(config)#monitor session 1 source interface GigabitEthernet 1/1 rx acl ruijie
                                   // 只监控 Gi0/1 口进入交换机并且只匹配 ACL 的流量
Ruijie(config)#monitor session 1 destination interface GigabitEthernet 1/24 switch
              // 设置接口镜像目的接口，后面加了一个关键字 switch，表示目的接口也能够上网
Ruijie(config)#end
```

【测试验证】

（1）在三层汇聚交换机上，使用"show monitor"命令查看接口镜像的状态。

```
Ruijie(config)#show monitor
sess-num: 1
span-type: LOCAL_SPAN
src-intf:
GigabitEthernet 0/1 frame-type Both
rx acl id 2900
```

任务 ㉛ 配置交换机基于流量的镜像安全

```
acl name ruijie
dest-intf:
GigabitEthernet 1/24
mtp_switch on              // 允许镜像目的接口转发数据流
```

（2）在三层汇聚交换机上查看 AC 信息。

```
Ruijie(config)#show access-lists
…       //  显示信息省略
```

（3）使用 Ethereal 等数据包捕获软件，确认流量是否只包含所需的流量。关于 Ethereal 等数据包捕获软件的应用，本书由于篇幅所限不予介绍。

任务 32 配置接入交换机保护端口

【任务描述】

某学校期末考试实施线上考试，为避免出现抄袭行为，希望机房所有计算机之间都不能互相访问，但能把考试结果提交给教师机。通过在机房接入交换机上实施保护端口技术，能实现上述效果。

交换机的保护端口技术，适用于同一台交换机下需要进行用户二层隔离的场景，如不允许同一个 VLAN 内的用户互访，必须完全隔离，可以有效防止病毒扩散攻击等。保护端口推荐在交换机每个端口接一个用户，这样才能实现基于端口的精确访问控制。

【任务目标】

配置交换机保护端口。

【组网拓扑】

按图 32-1 所示网络拓扑连接设备，配置交换机保护端口安全，PC1、PC2 属于 VLAN 10，PC3 属于 VLAN 20，PC1、PC2、PC3 之间不能相互访问，但是它们都能连接外网。

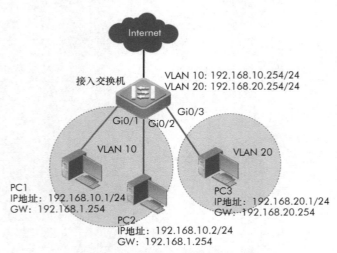

图 32-1 交换机保护端口网络拓扑

（备注：具体的设备连接接口信息可根据实际情况决定）

配置时需要注意如下事项。

（1）PC1 与 PC2 都属于 VLAN 10，可以通过配置保护端口来实现同网段之间的访问隔离，注意上连口不要开启。

任务 32 配置接入交换机保护端口

（2）PC3 与 PC1、PC2 属于不同 VLAN，在 PC3 所连接端口开启保护端口功能，并且全局开启路由阻断功能，即可实现不同网段之间的路由阻断。

（3）遇到接入交换机被下连一台集线器扩展并连接多个用户的情况时，交换机无法阻止集线器下的计算机互访。

【设备清单】

交换机（1台），网线（若干），PC（若干）。

【关键技术】

保护端口技术用来保护交换机接口之间的通信，当一个接口设为保护端口之后，保护端口之间无法通信，但保护端口与非保护端口之间可以正常通信。

保护端口有两种模式：一种是阻断保护端口之间的二层交换，但允许保护端口之间进行路由；另一种是同时阻断保护端口之间的二层交换和路由。

在两种模式都支持的情况下，第一种模式将作为默认配置模式。在 VSU 模式或 VSD 模式下，保护端口均有效。

【实施步骤】

```
Ruijie#configure terminal
Ruijie(config)#vlan 10
Ruijie(config-vlan)#vlan 20
Ruijie(config-vlan)#exit

Ruijie(config)#interface vlan 10
Ruijie(config-if-VLAN 10)#ip address 192.168.10.254 255.255.255.0
Ruijie(config-if-VLAN 10)#interface vlan 20
Ruijie(config-if-VLAN 20)#ip address 192.168.20.254 255.255.255.0
Ruijie(config-if-VLAN 20)#exit

Ruijie(config)#interface GigabitEthernet 0/1
Ruijie(config-if-GigabitEthernet 0/1)#switchport access vlan 10
Ruijie(config-if-GigabitEthernet 0/1)#switchport protected    // 开启保护端口
Ruijie(config-if-GigabitEthernet 0/1)#interface GigabitEthernet 0/2
Ruijie(config-if-GigabitEthernet 0/2)#switchport access vlan 10
Ruijie(config-if-GigabitEthernet 0/2)#switchport protected    // 开启保护端口
Ruijie(config-if-GigabitEthernet 0/2)#interface GigabitEthernet 0/3
Ruijie(config-if-GigabitEthernet 0/3)#switchport access vlan 20
Ruijie(config-if-GigabitEthernet 0/3)#switchport protected    // 开启保护端口
Ruijie(config-if-GigabitEthernet 0/3)#exit
Ruijie(config)#

Ruijie(config)#protected-ports route-deny
// 开启路由隔离功能，配置保护端口之间不能进行三层访问（可选）
/* 需要注意的是，目前仅部分产品支持和实现该功能，如锐捷网络 N18K/S86E/S78E 系列交换机 */
```

多层交换技术（实践篇）

【测试验证】

（1）使用 "show interface switchport" 命令，查看交换机接口配置信息，可以看到保护端口 Gi0/1-3 的状态是 Enabled，如图 32-2 所示。

```
Ruijie#show interface switchport
Interface              Switchport  Mode    Access  Native  Protected  VLAN
-------------------    ----------  ------  ------  ------  ---------  ----
GigabitEthernet 0/1    enabled     ACCESS  10      1       Enabled    ALL
GigabitEthernet 0/2    enabled     ACCESS  10      1       Enabled    ALL
GigabitEthernet 0/3    enabled     ACCESS  20      1       Enabled    ALL
GigabitEthernet 0/4    enabled     ACCESS  1       1       Disabled   ALL
GigabitEthernet 0/5    enabled     ACCESS  1       1       Disabled   ALL
GigabitEthernet 0/6    enabled     ACCESS  1       1       Disabled   ALL
```

图 32-2　查看交换机接口配置信息

（2）给网络中的测试计算机配置对应的 IP 地址信息，使用 Ping 命令，验证 PC1、PC2、PC3 之间的连通性。由于这些端口之间都实施了保护端口功能，因此无法 Ping 通，ARP 也无法学习到。

【注意事项】

（1）配置保护端口后，保护端口之间无法通信，但保护端口与非保护端口之间能正常通信。

（2）锐捷网络 N18K/S86E/S78E 系列交换机支持路由隔离功能。

（3）保护端口只能在同一台交换机上生效，如 PC1 属于 Switch A，PC2 属于 Switch B，Switch A 和 Switch B 都配置了保护端口功能，并且属于同网段，那么 PC1 和 PC2 之间依然可以互相访问。如果要实现跨交换机隔离，则需要使用 PVLAN 功能实现。

如图 32-3 所示场景，实施接入端口的保护端口功能，使用如下命令配置完成后，该交换机上实施了保护端口技术的 Fa0/2-24 上连接的设备之间都不能通信，但是所有端口都能和没有实施保护端口技术的 Fa0/1 上连接的服务器通信。

```
Switch(config)#interface range FastEthernet0/2-24
Switch(config-if-range)# switchport protected    // 配置端口为保护端口
```

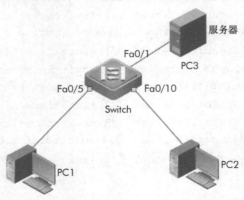

图 32-3　未保护端口和保护端口之间连通

任务 ㉝ Fat AP 组建会议室临时放装无线

【任务描述】

某公司会议室在早期建设过程中，只预留有一个有线接口，没有部署无线，日常部门开会过程中共享资源很不方便，故希望在会议室部署 Wi-Fi 无线环境，能接入公司办公网络，满足互联网的 Wi-Fi 无线访问需求，并实施 Wi-Fi 无线网络安全防护。

【任务目标】

配置 Fat AP 设备，组建临时放装模式的无线局域网。

【组网拓扑】

图 33-1 所示网络拓扑为某公司会议室内 WLAN 网络的部署场景，组建了临时放装无线办公网络。

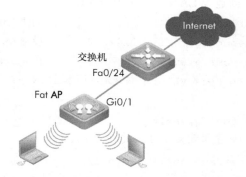

图 33-1　某公司会议室内 WLAN 网络的部署场景
（备注：具体的设备连接接口信息可根据实际情况决定）

【设备清单】

三层交换机（1 台），Fat AP（1 台），网线（若干），PC（若干）。

【关键技术】

无线接入点的 AP 设备俗称无线"热点"，可以扩大 Wi-Fi 无线信号覆盖范围，是 Wi-Fi 无线网络的组网核心。无线 AP 设备包括胖 AP（Fat AP）和瘦 AP（Fit AP）两种。

其中，Fat AP 模式是传统的 WLAN 组网设备，Fat AP 独自承担了认证终结、漫游切换、动态安全密钥产生等功能，因此称为胖 AP。

一般来说，Fat AP 可以单独组网，每一台 Fat AP 都需要配置才可以使用，但 Fat AP 设备配置复杂，且难以集中管理，建议在用户数量较少的组网场合使用。

多层交换技术（实践篇）

Fit AP 设备是大规模无线信号覆盖设备，Fit AP 设备自身不具有 WLAN 网络管理功能，需要和无线控制器组合使用，组建"Fit AP + AC"无线网络组网方案。

【实施步骤】

（1）组建会议室无线局域网。按图 33-1 所示网络拓扑连接设备，组建以 Fat AP 为中心的临时放装无线局域网。其中，POE 电源、Fat AP 设备采取直接的连接方式。

（2）切换 AP 模式为 Fat AP。默认情况下，断电后的 AP 为 Fit AP 模式。通过 AP 设备的 Console 口登录 AP 设备，在 AP 设备上切换 AP 的工作模式为 Fat AP 模式。

备注：登录 AP 设备时，如果提示输入密码，则默认密码为 ruijie（或 admin）。

```
Password: ruijie                    // 锐捷网络的设备默认密码为 ruijie
Ruijie>show ap-mode                 // 查看 AP 的当前模式
current mode: fit                   // AP 当前模式为 Fit AP（默认为瘦 AP 模式）

Ruijie>ap-mode fat                  // 修改 AP 的工作模式为 Fat AP（胖 AP 模式）
apmode will change to FAT......    // AP 设备重启，等待 1 分钟时间
```

（3）在 Fat AP 设备上创建用户 VLAN。

```
Ruijie(config)#
Ruijie(config)#vlan 10              // 创建用户 VLAN 10
```

（4）在 Fat AP 设备上创建 WLAN，关联用户 VLAN。

```
Ruijie(config)#dot11 wlan 1                     // 创建无线局域网 WLAN 1
Ruijie(dot11-wlan-config)#ssid ruijie10         // 设置 SSID 信息为 ruijie10
Ruijie(dot11-wlan-config)#vlan 10               // 在 WLAN 1 上关联 VLAN 10
/* 早期版本需要在 WLAN 下关联 VLAN；在锐捷最新版本 11.X 系统平台中，WLAN 下不再需要关联 VLAN */
Ruijie(dot11-wlan-config)#exit
```

（5）在 Fat AP 设备上配置射频接口 1，关联 WLAN。

```
Ruijie(config)#interface dot11 radio 1/0
Ruijie(config-if-Dot11radio 1/0)#encapsulation dot1Q 10
                    // 在射频接口上封装 Gi0/1 接口 DOT1Q 协议，并映射给 VLAN 10
Ruijie(config-if-Dot11radio 1/0)#wlan 1         // 该射频接口和 WLAN 1 关联
Ruijie(config-if-Dot11radio 1/0)#exit
```

（6）在 Fat AP 设备上配置默认网关。

```
Ruijie(config)#ip route 0.0.0.0  0.0.0.0 172.16.10.1
                                                // 配置 AP 的默认网关（路由）
```

（7）在三层交换机上，配置三层交换机用户 VLAN，分配地址。

```
Switch(config)#
Switch(config)#vlan 10              // 创建用户 VLAN
Switch(config-vlan)#exit
Switch(config)#interface vlan 10    // 打开 VLAN 10 接口
Switch(config-if-vlan 10)#ip address 172.16.10.1 255.255.255.0
```

任务 ㉝ Fat AP 组建会议室临时放装无线

```
                                                        // 配置网关地址
Switch(config-if-vlan 10)#exit

Switch(config)#service dhcp        // 给用户 VLAN 分配地址
Switch(config)#ip dhcp pool USE10-IP    // 自动获取地址池名称 USE10-IP
Switch(dhcp-config)#network 172.16.10.0 255.255.255.0
Switch(dhcp-config)#default-router 172.16.10.1    // 设置默认网关


Switch(config)#int FastEthernet 0/24
Switch(config-if-FastEthernet 0/24)#switch access vlan 10
              // 该接口分配到 VLAN 10 中,实现 VLAN 10 连接的一台 AP 设备
Switch(config-if-FastEthernet 0/24)#exit
```

（8）在 Fat AP 设备上实施 WLAN 无线安全防范。

```
Ruijie(config)#
Ruijie(config)#wlansec 1    // 进入 WLAN 1 的无线安全模式
Ruijie(config-wlansec)#security wpa enable
              // 配置 WPA 或 RSN(WPA2) 认证模式
Ruijie(config-wlansec)#security wpa ciphers aes enable
       // 配置 WPA 认证数据加密模式（有 TKIP 和 AES 两种模式）为 AES
Ruijie(config-wlansec)#security wpa akm psk enable
              //  配置 WPA 接入认证方式为 PSK
Ruijie(config-wlansec)#security wpa akm psk set-key ascii 12345678
              // 配置 PSK 密码为 12345678
```

【注意事项】

（1）使用 WPA/RSN 认证时，需要配置数据加密方式和接入认证方式。

（2）如果配置接入认证方式为 PSK，则需配置 PSK 密码。

（3）在同一个 WLAN 安全模式下，WPA/RSN 认证模式不能与 WEP 模式同时配置。

（4）在智能手机或者便携式计算机上，测试和验证 WLAN 无线信号。

打开手机，搜索附近的 Wi-Fi 无线信号，可以收到配置的 SSID 符号，查看手机获取到的 IP 地址，接入无线局域网。

任务 ㉞ 组建单核心无线校园网（Fit AP+AC）

【任务描述】

某学院校园网实施无线校园网建设项目，希望实现更广的无线网络信号覆盖。如果在 Wi-Fi 无线网络中部署的 AP 数量众多，就需要进行统一管理，Fat AP 不能适应新的组网要求，需要增购一台无线控制器 AC 设备，改造为"Fit AP+AC"模式的无线校园网。

【任务目标】

配置 AC 设备，组建"Fit AP+AC"智能无线校园网。

【组网拓扑】

图 34-1 所示网络拓扑为某校园网 Wi-Fi 无线信号覆盖的部分场景，使用 Fit AP 直接连接 AC，组建单核心无线校园网。

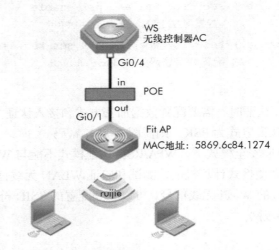

图 34-1 某校园网 Wi-Fi 无线信号覆盖的部分场景
（备注：具体的设备连接接口信息可根据实际情况决定）

【设备清单】

Fit AP（1台），无线控制器 AC（1台），AP 供电模块 E130（1套），测试便携式计算机（两台），网线（若干）。

【关键技术】

Fat AP 是具有独立组网功能的无线局域网组网设备，能实现 WLAN 网络中物理层上的

任务 34　组建单核心无线校园网（Fit AP+AC）

用户数据加密；实现用户接入安全认证；提供 QoS 服务管理，实现网络安全管理；提供无线漫游等无线网络运维和管理功能。但 Fat AP 设备难以实现智能化无线局域网的集中管理，只适用于小型无线局域网组网场景。

在大型无线校园网组网中，需要使用 Fit AP 进行更大范围的无线网络信号的覆盖。Fit AP 由于功能单一，不能独立工作，但可以通过无线控制器 AC 管理，完成无线网络中的信号接入、无线网络控制和管理。

在"Fit AP +AC"的无线组网方案中，AP 上线"零配置"，所有配置都集中到 AC 上实施完成，以便于集中管理，能实现三层漫游等 Fat AP 不具备的功能。

以"Fit AP+AC"为核心的智能无线网络管理方案中，无线客户端和 AP 的关联时间大大减少，可以实现如 PDA、手持终端、便携式计算机等无线终端设备在校园无线网络里快速切换，实现漫游的功能。

【实施步骤】

（1）组建学校办公楼无线局域网。按照图 34-1 所示网络拓扑连接设备，组建以 AC 为中心的 Fit AP 模式无线局域网。其中，POE 电源、Fit AP 设备采用直接的连接方式。

（2）切换 AP 设备的模式为 Fit AP。默认情况下，AP 的模式为 Fit AP 模式。可通过 Console 口登录 AP，在 AP 上切换其模式为 Fit AP。

备注：登录 AP 设备时，如果提示输入密码，则默认密码为 ruijie（或 admin）。

```
Password: ruijie
```

使用以下命令确认 Fit AP 模式。

```
Ruijie>show ap-mode                // 查看 AP 的当前模式
current mode: FIT                  // AP 当前模式为 Fit AP
Ruijie#configure terminal
Ruijie(config)#ap-mode FIT
          // 修改成瘦 AP（Fit AP）模式，胖瘦模式切换后设备会重启，请耐心等待 1 分钟左右
```

（3）在 AC 设备上，创建用户 VLAN、Fit AP 的 VLAN。

```
Ruijie#configure terminal
Ruijie(config)#vlan 10              // 创建 Fit AP 的 VLAN 10
Ruijie(config-vlan)#vlan 20         // 创建用户的 VLAN 20
```

（4）在 AC 设备上，配置 Fit AP、无线用户网关和 Loopback0 地址。

```
Ruijie(config)#interface vlan 10    // 配置 AP 的网关
Ruijie(config-if-vlan 10)#ip address 172.16.1.1 255.255.255.0
Ruijie(config-if-vlan 10)#no shutdown
Ruijie(config-if-vlan 10)#interface vlan 20      // 开启用户的 SVI
Ruijie(config-if-vlan 20)#ip address  172.17.1.1 255.255.255.0
//配置用户的默认网关
Ruijie(config-if-vlan 20)#no shutdown

Ruijie(config-int-vlan)#interface loopback 0    //  开启 Loopback0 接口
Ruijie(config-int-loopback 0)#ip address 1.1.1.1 255.255.255.255
```

多层交换技术（实践篇）

```
                // 配置 Loopback0 地址作为 AC 地址，用于 AP 寻找 AC，即 DHCP 中的 option 138 字段
    Ruijie(config-int-loopback 0)#exit
```

（5）在 AC 设备上，配置 Fit AP 和 AC 之间的通信连接通道。

```
    Ruijie(config)#
    Ruijie(config)#interface GigabitEthernet 0/4
    Ruijie(config-int-GigabitEthernet 0/4)#switchport access vlan 10
                // 配置 AC 与 Fit AP 相连的接口，把接口划到 Fit AP 所属的 VLAN 中
    Ruijie(config-int-GigabitEthernet 0/4)#exit
```

（6）在 AC 设备上，配置 Fit AP 设备的 DHCP，分配 AP 设备的地址。

```
    Ruijie(config)#
    Ruijie(config)#service dhcp              // 开启 DHCP 服务
    Ruijie(config)#ip dhcp pool ap-ruijie    // 创建 Fit AP 的地址池，名称是 ap-ruijie
    Ruijie(config-dhcp)#option 138 ip 1.1.1.1
                // 配置 option 字段，指定 AC 的地址，即 AC 的 Loopback0 地址
    Ruijie(config-dhcp)#network 172.16.1.0 255.255.255.0   // 分配给 Fit AP 的地址范围
    Ruijie(config-dhcp)#default-route 172.16.1.1           // 分配给 Fit AP 的网关
```

注意：AP 的 DHCP 中的 option 字段和网段、网关要配置正确，否则会出现 AP 获取不到 DHCP 信息，导致无法建立隧道的情况。

（7）在 AC 设备上，配置用户 DHCP，分配给用户 IP 地址。

```
    Ruijie(config)#
    Ruijie(config)#service dhcp         //开启 DHCP 服务
    Ruijie(config)#ip dhcp pool user-ruijie
                // 配置用户 DHCP 地址池，名称是 user-ruijie
    Ruijie(config-dhcp)#network 172.17.1.0 255.255.255.0
    // 无线网络中用户自动获取的地址范围
    Ruijie(config-dhcp)#default-route 172.17.1.1     // 配置无线网络中用户的网关
    Ruijie(config-dhcp)#exit
```

（8）在 AC 设备上，查看 Fit AP 设备正常获取的 IP 地址信息。

```
    Ruijie#Show ip dhcp  binding    // 查看 Fit AP 设备获取的 IP 地址
    …
```

（9）在 AC 设备上，创建 WLAN 信息。

```
    Ruijie(config)#wlan-config 1  ruijie-2
      // 配置 WLAN-config 信息，WLAN 的识别号是 WLAN 1，其无线信号 SSID 是 ruijie-2
    Ruijie(config-wlan)#exit
```

（10）在 AC 设备上，创建 Fit AP 组，关联 WLAN-config 和用户 VLAN。

```
    Ruijie(config)#ap-group abc
                // 使用 ap-group 命令创建 Fit AP 组，开启 Fit AP 组配置功能
    Ruijie(config-ap-group)#interface-mapping 1 20
                // 对 WLAN-config 1 和用户 VLAN 20 进行关联
    Ruijie(config-ap-group)#end
```

任务 ㉞　组建单核心无线校园网（Fit AP+AC）

（11）在 AC 设备上，查看关联的 WLAN-config 和用户 VLAN。

```
Ruijie#Show ap-config summary      // 查看 Fit AP 的配置信息
    ...

Ruijie#Show capwap state           // 查看 Fit AP 和 AC 之间的隧道能否正常建立
    ...
```

（12）在 AC 设备上，把 AC 上的配置下发到指定 MAC 地址的 Fit AP 设备中。

```
Ruijie(config)#ap-config  5869.6c84.1274
                                   // 这里 5869.6c84.1274 为指定 Fit AP 的 MAC 地址
                                   // 该命令把 AP 组配置关联到指定 Fit AP。在该 Fit AP 上应用 ap-group 组
                                   // 第一次部署时，默认 5869.6c84.1274 是 AP 的 MAC 地址
Ruijie(config-ap-config)#ap-group abc
                                   // 把 AP 组配置关联到指定 Fit AP 上
Ruijie(config-ap-group)#end
```

注意：要正确配置 ap-group 命令，否则会有无线用户搜索不到 SSID 的情况发生。

【测试验证】

（1）在 AC 上验证配置命令。

```
Ruijie#show  running-config        // 查看 AC 设备的配置信息
    ...


Ruijie#show ap-config  summary     // 查看 AP 设备上的配置信息
    Online AP number: 1
    Offline AP number: 0
AP Name       IP Address    Mac Address   Radio  Radio  Up/Off  time  State
-----------------------  ----------  -------  ---------------  -------  -----
5869.6c84.1274 172.16.1.2 5869.6c84.1274 1 E 2 11 100 2 E 1 149* 100 0:00:20:02 Run


Ruijie#show ac-config client       // 查看 AC 设备上连接的客户端信息
AP    : ap name/radio id
Status: Speed/Power Save/Work Mode/Roaming State, E = enable power save,
D = disable power save
Total StaNum : 2
    STA MAC   IPV4 Address    AP   WlanVlan Status  AssoAuth  Net Auth  Up time
    -------   ---------------  ----  ----  ------  -----  --------  -------
    a08d.160e.7964   172.17.1.3   5869.6c84.1274/1   2   2   52.0M/E/bgn   OPEN  OPEN   0:00:24:48
    b809.8a74.c783   172.17.1.5   5869.6c84.1274/2   2   2   72.5M/D/an    OPEN  OPEN   0:00:24:06


Ruijie#show  capwap  state         // 查看 AC 设备和 Fit AP 的隧道信息
    CAPWAP tunnel state, 1 peers, 1 is run:
```

多层交换技术（实践篇）

```
Index      Peer IP                 Port        State       Mac Address
1          172.16.1.2              10000       Run         5869.6c84.1274

Ruijie#show ip dhcp binding         // 查看 Fit AP 的地址信息
Total number of clients     : 7
Expired clients             : 1
Running clients             : 6
--------------------------------------------------------------
IP address        Hardware address      Lease expiration            Type
172.16.1.2        5869.6c84.1275        000 days 23 hours 23 mins   Automatic
172.17.1.4        e850.8b69.7907        000 days 23 hours 25 mins   Automatic
172.17.1.3        a08d.160e.7964        000 days 23 hours 25 mins   Automatic
172.17.1.2        28e3.1f76.afa1        000 days 23 hours 55 mins   Automatic
172.17.1.5        b809.8a74.c783        000 days 23 hours 26 mins   Automatic
```

```
AC#show  running-config         // 查看 AC 设备的配置信息
...
```

```
AC#show ap-config  summary      // 查看 AP 设备的配置信息
...
```

```
AC#show  ac-config client       // 查看关联的终端信息
...
```

```
AC#show  capwap  state          // 查看 AC 设备和 Fit AP 的隧道信息
...
```

（2）查看无线网络连接信息（测试便携式计算机）。
打开测试便携式计算机，查看无线网络连接信息，图 34-2 表示无线网络连接成功。

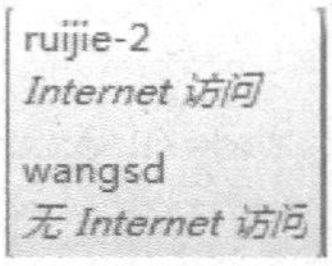

ruijie-2
Internet 访问
wangsd
无 Internet 访问

图 34-2　无线网络连接成功

（3）查看无线网卡获取到的 IP 地址信息（测试便携式计算机）。
在测试便携式计算机上进入 DOS 命令行，使用"ipconfig"命令查看无线终端计算机设备成功获取到的 IP 地址。

任务 35 组建单核心无线校园网（Fit AP+SW+AC）

【任务描述】

"Fit AP+AC"的智能无线组网方案极大地扩展了WLAN无线网络中信号的覆盖范围，实现了全区域的随时、随地、方便的移动网络接入方案。"Fit AP+AC"组网需要依托有线核心网络实施，因此，AC经常不和AP直连，多直接连接在有线网络三层核心交换机上，通过配置隧道技术，实现Fit AP和AC通信。

【任务目标】

组建"Fit AP+SW+AC"模式智能无线局域网。

【组网拓扑】

图35-1所示网络拓扑为某大学无线校园网场景，Fit AP通过交换机连接AC组建智能无线，组建单核心的无线校园网。

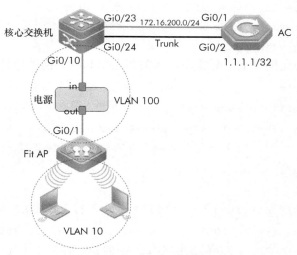

图 35-1　某大学无线校园网场景

（备注：具体的设备连接接口信息可根据实际情况决定）

【设备清单】

Fit AP（1台），无线控制器AC（1台），AP供电模块E130（1套），三层交换机（1台），测试便携式计算机（两台）。

多层交换技术（实践篇）

【实施步骤】

（1）组建校园中某行政办公楼无线局域网。

按图 35-1 所示网络拓扑连接设备，组建以 AC 为中心的智能无线局域网，需重点注意 POE 电源、Fit AP 设备、AC 和三层核心交换机的连接方式。

（2）切换 AP 模式为 Fit AP。默认情况下，AP 的模式为 Fit AP 模式。可通过 AP 的 Console 口登录 AP，在 AP 上查看 AP 的模式，否则需要切换模式为 Fit AP。

登录 AP 设备时，如果提示输入密码，可默认密码为 ruijie（或 admin）。

```
Password: ruijie
```

或者使用如下命令查看 AP 模式。

```
Ruijie>show ap-mode                // 查看 AP 的当前模式
current mode: FIT                  // AP 当前模式为 Fit AP（默认）
Ruijie#configure terminal
Ruijie(config)#ap-mode FIT
            //修改成 Fit AP 模式，胖瘦模式切换后设备会重启，需耐心等待 1 分钟左右
```

（3）在三层核心交换机上，创建用户 VLAN 和网关 VLAN（有线）。

```
Ruijie#configure terminal
Ruijie(config)#hostname Switch
Switch(config)#vlan 10                  // 创建用户 VLAN 10
Switch(config-vlan)#name user           // 命名 VLAN 10 为用户区域

Switch(config-vlan)#vlan 100            // 创建 AP 的 VLAN
Switch(config-vlan)#name AP             // 命名 VLAN 100 为 AP 的网关
Switch(config-vlan)#exit
```

（4）在三层核心交换机上，配置用户网关和无线 Fit AP 网关地址。

```
Switch(config)#
Switch(config)#interface vlan 10        // 进入用户的 VLAN
Switch(config-if-vlan 10)#ip address 172.16.10.1  255.255.255.0
            // 配置用户 VLAN 10 的 SVI 接口 IP 地址，作为无线用户的网关地址
Switch(config-if-vlan 10)#exit

Switch(config)#interface vlan 100       // 打开 Fit AP 的 VLAN
Switch(config-if-vlan 100)#ip address 172.16.100.1 255.255.255.0
            // 配置用户 Fit AP 的 SVI 接口 IP 地址，作为 Fit AP 网关地址
Switch(config-if-vlan 10)#exit
```

（5）在三层核心交换机上，配置连接 AC 和 AP 的接口链路。

① 配置连接 AP 的接口，把 AP 接入有线网络。

```
Switch(config)#
Switch(config)#interface GigabitEthernet 0/10    // 打开连接 Fit AP 的接口
Switch(config-if-GigabitEthernet 0/10)#switchport access vlan 100
```

任务 ㉟ 组建单核心无线校园网（Fit AP+SW+AC）

```
                                    // 连接AP设备的接口划分给AP所在的VLAN 100
Switch(config-if-GigabitEthernet 0/10)#exit
```

② 配置连接AC的接口23为三层路由接口，实现AP三层路由可达AC。

```
Switch(config)#interface GigabitEthernet 0/23    // 进入连接AC的接口
Switch(config-if-GigabitEthernet 0/23)#no switchport
                // 配置连接AC的接口为三层路由接口，实现AP三层IP路由可达AC
Switch(config-if-GigabitEthernet 0/23)#ip address 172.16.200.1 255.255.255.0
Switch(config-if-GigabitEthernet 0/23)#exit
```

③ 配置连接AC的接口24为二层干道接口，实现AP上的二层交换数据可达AC。

```
Switch(config)#interface GigabitEthernet 0/24    // 进入连接AC的接口
Switch(config-if-GigabitEthernet 0/24)#switchport mode trunk
                // 配置连接AC的另一接口为干道接口，实现Fit AP和AC的二层通信
Switch(config-if-GigabitEthernet 0/24)#exit
```

（6）在三层核心交换机上，配置用户DHCP服务，给无线用户分配IP地址。

```
Switch(config)#service dhcp                  // 开启DHCP服务
Switch(dhcp-config)#ip dhcp pool user-ip     // 配置无线用户地址池
Switch(dhcp-config)#network 172.16.10.0 255.255.255.0
                                             // 给无线用户配置自动获取地址
Switch(dhcp-config)#default-router 172.16.10.1    // 配置用户设备默认网关
Switch(dhcp-config)#exit
```

（7）在三层核心交换机上，配置无线Fit AP的DHCP服务，给无线接入AP设备分配IP地址。

```
Switch(config)#service dhcp                  // 开启DHCP服务
Switch(dhcp-config)#ip dhcp pool fit-ap-ip   // 给Fit AP分配地址池
Switch(dhcp-config)#network 172.16.100.0 255.255.255.0
                                             // 给Fit AP自动分配地址
Switch(dhcp-config)#default-router 172.16.100.1    // 配置Fit AP默认网关
Switch(dhcp-config)#option 138 ip 1.1.1.1
// 配置DHCP option选项，分配1.1.1.1地址给AC，帮助Fit AP设备在网络中找到AC
Switch(dhcp-config)#exit
```

（8）在三层核心交换机上，配置到达AC的静态路由（有线）。

```
Switch(config)#
Switch(config)#ip route 1.1.1.1 255.255.255.255 172.16.200.2
        // 配置三层核心交换机设备指向AC路由，实现AC和交换机的互连互通
```

（9）在AC设备上，配置WLAN无线信息。

① 给AC设备命名。

```
Ruijie(config)#hostname AC      // 修改AC的名称
AC(config)#
```

多层交换技术（实践篇）

② 配置 AC 连接交换机的三层接口，实现路由可达。

```
AC(config)#interface GigabitEthernet 0/1        // 进入连接三层交换机接口
AC(config-if-GigabitEthernet 0/1)#no switchport
AC(config-if-GigabitEthernet 0/1)#ip address 172.16.200.2 255.255.255.0
            // 配置连接三层核心交换机的三层接口的地址，实现 AC 和交换机三层路由的互连
AC(config-if-GigabitEthernet 0/1)#exit
```

③ 配置 AC 连接交换机的二层接口。

```
AC(config)# interface GigabitEthernet 0/2        // 进入连接三层交换机接口
AC(config-if-GigabitEthernet 0/2)#switchport mode trunk
            // 配置连接三层核心交换机的另一接口为干道接口，实现 Fit AP 通信
AC(config-if-GigabitEthernet 0/2)#exit
```

（10）在 AC 上，配置指向三层核心交换机的路由。

```
AC(config)#interface Loopback 0
// 必须是 Loopback 0，用于 Fit AP 寻找 AC 的 IP 地址中 DHCP 的 option 138 字段
AC(config-if-Loopback 0)#ip address 1.1.1.1 255.255.255.255
AC(config-if-Loopback 0)#exit
AC(config)#ip route 0.0.0.0 0.0.0.0 172.16.200.1
            // 建立 AC 和三层核心交换机之间的默认路由，172.16.200.1 是下一跳地址
```

（11）在 AC 上，创建 VLAN 无线信息。

① 在 AC 上创建 VLAN 信息。

```
AC(config)#vlan 10          // 用户的 VLAN
AC(config-vlan)#vlan 100        // Fit AP 的 VLAN（可选）
AC(config-vlan)#exit
```

② 激活创建的 VLAN（可选）。

```
AC(config)#int vlan 10        // 激活用户 VLAN，激活用户 SVI 接口（必须配置）
AC(config-if-vlan 10)#int vlan 100      // 激活 Fit AP 的 VLAN
AC(config-if-vlan100)#exit
```

（12）在 AC 设备上，创建 WLAN 无线信息。

```
AC(config)#wlan-config 1 ruijie-2    // 创建 WLAN 的 ID 为 1，SSID 为 ruijie-2
AC(config-wlan)#exit
```

（13）在 AC 设备上，创建 AP 组，关联 WLAN-config 和用户 VLAN（无线）。

```
AC(config)#ap-group ab          // 创建 AP 组 ab，开启 ab 组配置
AC(config-ap-group)#interface-mapping 1 10
// 将用户 VLAN 10 和 WLAN 1 建立关联
AC(config-ap-group)#end
```

（14）在 AC 上查看关联 WLAN-config 和隧道信息（无线）。

```
AC#Show ap-config summary      // 查看 Fit AP 的配置信息
...
```

任务 35 组建单核心无线校园网（Fit AP+SW+AC）

```
AC#Show capwap state        // 查看 Fit AP 和 AC 之间的隧道信息
...
```

（15）把 AC 上的配置分配到 Fit AP 设备（无线）中。

```
AC(config)#ap-config  AP-MAC
// 第一次部署，默认 AP 的 MAC 地址，打开 AP1（已命名 ap，输入 AP 名称，如果没有，需要创建）
AC(config-ap-config)#ap-group ab           // 把 AP 加入新建的 ab 组
AC(config-ap-group)#end
```

注意：要正确配置 ap-group，否则会有无线用户搜索不到 SSID 的情况发生。

【测试验证】

（1）在 AC 上验证配置命令（无线）。

```
AC#show  running-config     // 查看 AC 设备的配置信息
...
```

```
AC#show ap-config  summary   // 查看 AC 设备的配置信息
...
```

```
AC#show ac-config client     // 查看关联的终端信息
...
```

```
AC#show  capwap  state       // 查看 AC 设备和 Fit AP 的隧道信息
...
```

（2）查看无线网络连接信息（测试便携式计算机）。

打开测试便携式计算机，查看无线网络连接信息，如图 35-2 所示，表示无线网络连接成功。

（3）查看无线网卡获取到的 IP 地址信息（测试便携式计算机）。

在测试便携式计算机上进入 DOS 命令行，使用 "ipconfig" 命令查看无线终端计算机设备成功获取到的 IP 地址信息，如图 35-3 所示。

图 35-2 无线网络连接成功   图 35-3 测试便携式计算机成功获取到的 IP 地址

【备注】

为减少连线并优化网络，在 AC 和三层核心交换机之间经常使用一根网线连接，如图 35-4 所示，AC 和三层核心交换机之间通过 SVI 接口实现三层连通。此时，需要在三层核心交换机上删除上述的 Gi0/23 口上的三层路由配置，增加如下配置，AC 上也是如此。

多层交换技术（实践篇）

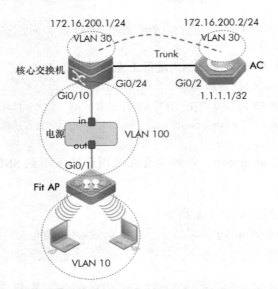

图 35-4　AC 和三层核心交换机之间通过 SVI 接口实现三层连通

（1）需要在三层核心交换机上增加如下配置。

① 创建 VLAN 30，配置连接 AC 的 SVI 接口 IP 地址。

```
Switch(config)#vlan 30
Switch(config-vlan)#interface vlan 30
Switch(config-if-vlan 30)#ip address 172.16.200.1 255.255.255.0
Switch(config-if-vlan 30)#no shutdown
Switch(config-if-vlan 30)#exit
```

② 配置默认路由，指向 AC 设备，配置连接 AC 的 SVI 接口 IP 地址。

```
Switch(config)# ip route 1.1.1.1 255.255.255.255 172.16.200.2
```

（2）需要在 AC 设备上增加如下配置。

① 创建 VLAN 30，配置连接三层核心交换机的 SVI 接口 IP 地址。

```
AC(config)#vlan 30
AC(config-vlan)#interface vlan 30
AC(config-if-vlan 30)#ip address 172.16.200.2 255.255.255.0
AC(config-if-vlan 30)#no shutdown
AC(config-if-vlan 30)#exit
```

② 配置默认路由，指向核心交换机设备，配置连接核心交换机的 SVI 接口 IP 地址。

```
AC(config)#ip route 0.0.0.0 0.0.0.0 172.16.200.1
           // 建立 AC 和三层核心交换机之间的默认路由，172.16.200.1 是下一跳地址
```

任务 36 综合实训 1

【任务描述】

中亚电子商务公司总部在福州市,由于业务发展,在泉州和厦门设置了两个分公司,构建了一个跨越三地的集团网络。福州总部有销售部、营销部、市场部和管理部 4 个部门。福州总部网络使用专线接入互联网,分别实现和厦门、泉州分公司的网络连接。

厦门分公司使用出口路由器接入互联网,其内部办公网采用全无线架构,使用"AC+Fit AP"模式实现办公无线覆盖。泉州分公司通过出口路由器接入互联网,办公网内部通过单臂路由方式,实现办公网连通。

【组网拓扑】

中亚电子商务公司总部连接泉州和厦门两个分公司的网络拓扑如图 36-1 所示,相关接口及对应的 VLAN 信息如下。

VLAN 10:总部销售部,规划接口有 Fa0/1-5。
VLAN 20:总部营销部,规划接口有 Fa0/6-10。
VLAN 30:总部市场部,规划接口有 Fa0/11-15。
VLAN 40:总部管理部,规划接口有 Fa0/16-20。
VLAN 50:总部服务器区,规划接口有 Fa0/2-6。
VLAN 60:泉州分公司管理部,规划接口有 Fa0/2-10 。
VLAN 70:泉州分公司销售部,规划接口有 Fa0/11-20。
VLAN 80:厦门分公司办公网无线 AP VLAN。
VLAN 90:厦门分公司办公网无线用户 VLAN。

【设备清单】

根据表 36-1,使用相应线缆将所有设备连接起来。

多层交换技术（实践篇）

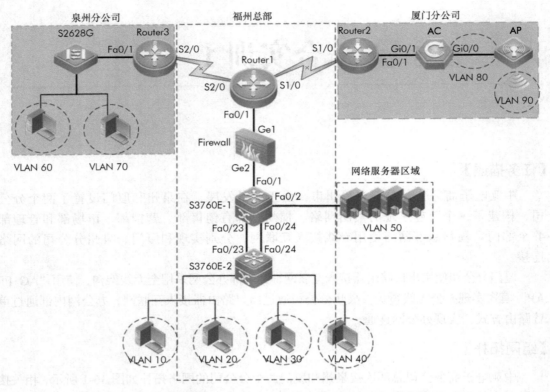

图 36-1　中亚电子商务公司总部连接泉州和厦门两个分公司的网络拓扑

表 36-1　线缆连接及 IP 地址规划

设备	设备名称	设备接口	IP 地址	备注
路由器	Router1（RSR20-1）	Fa0/1	10.0.1.1/24	总部内网接口
		S1/0	10.0.2.1/24	通过互联网连接厦门分公司
		S2/0	10.0.3.1/24	通过互联网连接泉州分公司
	Router2（RSR20-2）	Fa0/1	172.16.1.1/24	厦门分公司内网接口
		S1/0	10.0.2.2/24	通过互联网连接公司总部
	Router3（RSR20-3）	S2/0	10.0.3.2/24	通过互联网连接公司总部
		Fa0/1	10.0.4.1/24	泉州分公司内网接口
防火墙	Firewall	Ge1	#	网桥模式，连接出口路由器
		Ge2	#	网桥模式，连接核心交换机
三层交换机	S3760E-1	Fa0/1	10.0.1.2/24	连接防火墙接口
		Fa0/2-6	#	总部服务器接口
		Fa0/23	#	连接核心交换机接口

任务 36 综合实训 1

续表

设备	设备名称	设备接口	IP 地址	备注
三层交换机	S3760E-1	Fa0/24	#	连接核心设备接口
		VLAN 10 SVI（销售部）	10.0.10.1/24	销售部 SVI 虚拟网关
		VLAN 20 SVI（营销部）	10.0.20.1/24	营销部 SVI 虚拟网关
		VLAN 30 SVI（市场部）	10.0.30.1/24	市场部 SVI 虚拟网关
		VLAN 40 SVI（管理部）	10.0.40.1/24	管理部 SVI 虚拟网关
	S3760E-2	Fa0/23	#	连接核心设备接口
		Fa0/24	#	连接核心设备接口
		Fa0/1-5	#	销售部接入设备接口
		Fa0/6-10	#	营销部接入设备接口
		Fa0/11-15	#	市场部接入设备接口
		Fa0/16-20	#	管理部接入设备接口
接入交换机	S2628G	Fa0/2-10	#	泉州分公司管理部接口
		Fa0/11-20	#	泉州分公司销售部接口
无线设备	AC	Gi0/1	172.16.1.2/24	厦门分公司出口路由器接口
		AC 环回接口地址	9.9.9.9/24	AC 环回接口
	FIT-AP	VLAN 80	10.0.80.1/24	Fit AP 的设备 VLAN
		VLAN 90	10.0.90.1/24	用户的 VLAN
备注	路由器 Serial 接口连接标识应尽量保持一致，如果由于实验的设备硬件缺少接口，造成无法和文档保持一致，应在文档前面予以说明			

福州总部办公网中搭建了多台网络服务器，具体应用如图 36-1 所示，根据表 36-2 列出的功能配置网络服务器应用系统。

表 36-2 网络服务器地址规划及业务信息

宿主机	虚拟服务器名称	提供服务	操作系统	IP 地址
第 1 台计算机（PC1）	dc.zxb.com	域服务/DNS 服务 证书服务	Windows Server 2008	10.0.5.8/24
	ftp.zxb.com	FTP 服务	CentOS 6.5	10.0.5.14/24～10.0.5.16/24

多层交换技术（实践篇）

续表

宿主机	虚拟服务器名称	提供服务	操作系统	IP 地址
第 2 台计算机（PC2）	www.zxb.com	Web 服务	Windows Server 2008	10.0.5.12/24
	dhcp.zxb.com	DHCP 服务 NFS 服务	CentOS 6.5	10.0.5.11/24
第 3 台计算机（PC3）	mail.zxb.com	邮件服务	Windows Server 2008 R2	10.0.5.10/24
	bdns.zxb.com	备份 DNS 服务	Windows Server 2008 CORE	10.0.5.9/24
第 4 台计算机（PC4）	pc1.zxb.com	测试	Windows 7	DHCP
	pc2.zxb.com	测试	Windows 7	DHCP
备注	任务实施过程中，根据现场提供的计算机设备数量，可以将相关的服务内容合并配置在一台计算机设备上			

【任务目标】

按照拓扑要求，构建一个跨越三地的集团网络，完成如下网络构建需求。

1. 网络搭建

（1）按照图 36-1 所示信息，给所有设备命名，配置所有设备接口、子接口、VLAN 接口上的 IP 地址，使所有互连设备能正常通信。

（2）根据拓扑图，在总部的核心交换机上创建相应 VLAN，每个 VLAN 使用部门名称拼音命名，如销售部为"xiaoshoubu"，其他以此类推。

（3）在总部交换机上完成 VLAN 创建后，根据表 36-1 网络规划信息将交换机相关接口划分到相应 VLAN 中。

（4）对总部路由器与分公司连接的接口增加描述信息，如对于 Router1 的 S2/0 接口与泉州分公司 Router3 的 S2/0 接口连接，其相应描述信息为"Router1-TO-Router2 interface-S2/0"，以便今后查找，其他以此类推。

（5）总部三层核心交换机与核心交换机之间使用双链路连接，形成冗余链路。配置 MSTP 创建两个生成树实例，分别为实例 10 和实例 20。将 VLAN 10 和 VLAN 20 加入实例 10，将 VLAN 30 和 VLAN 40 加入实例 20。使 VALN 10、VLAN 20 中数据流量通过第一条链路传输，VALN 30、VLAN 40 数据流量通过第二条链路传输，以保障网络稳健，同时实现均衡负载。

（6）在总部的网络中，为了防止网络 DHCP 攻击，需要核心交换机和三层核心交换机上部署 DHCP 监听，保障所有 VLAN 中的客户机都能正常获取 IP 地址。在总部三层核心交换机上配置总部网络中的 DHCP 服务器，保障所有 VLAN 设备自动获取本网 IP 地址。

（7）总部与厦门分公司之间的广域网连接租用电信专线链路传输，为保证数据在公网上传输安全，需要在总部和两家分公司的接入路由器上使用 IPSec 技术对数据加密。其中，VPN 采用隧道模式，预共享密码为 123456，数据采用 ESP-DES、ESP-HASH-MD5 加密方式，AH 采用 Hash 方式加密。

（8）为保障总部与泉州分公司之间的广域网连接链路安全，需要在电信专线链路上配置 PPP 协议，并开启 CHAP 安全认证，要求以总部为验证端，分公司为被验证端，口令为 ruijie。

（9）全网配置 OSPF 动态路由实现连通。其中，总部网络设备 Router1 的 RID 为 1.1.1.1；Router2 的 RID 为 2.2.2.2；Router3 的 RID 为 3.3.3.3；S3760E-1 的 RID 为 4.4.4.4。

（10）为保障 OSPF 路由安全，在路由更新时采用基于接口 MD5 验证方式，口令为 ruijie。

（11）厦门分公司使用 NAT 技术，把内网中的私有地址转换为路由器外接口（全局地址）。允许内网用户在上班时间（周一至周五 9:00～18:00）访问互联网。

（12）在厦门分公司路由器配置默认路由，保障网络连通，并使用路由重分发的技术，将默认路由发布到 OSPF 网络中，网络类型为 E1，开销为 220。

（13）泉州分公司租用电信专线链路与总部相连，其内网采用二层交换机做办公网接入，由于内网有两个部门 VLAN，所以需要使用单臂路由实现 VLAN 间路由的连通。

（14）在总部出口区部署防火墙，保障服务器区域的安全性。同时，为保障总部网络的传输速度，配置防火墙为网桥模式。

（15）在总部的防火墙上配置防火墙策略，允许总部内网中的用户（VLAN 10、VLAN 20、VLAN 30、VLAN 40）每天 9:00～18:00 访问外部互联网，允许总部网络中服务器全天对外提供服务。

（16）在总部出口路由器上配置 NAT 技术，将服务器区中的 Web 服务和 FTP 服务发布到互联网，其合法地址为总部出口路由器 Router1 设备外网口 Fa0/2 接口地址。

（17）配置厦门分公司"AC+Fit AP"结构的无线局域网环境，并能够访问互联网。其中，厦门分公司在无线交换机上安装 DHCP 服务器，为内部无线用户动态分配 IP 地址、网关和 DNS 服务器（218.85.152.99）；其分配地址段为 10.0.9.10～10.0.9.200。厦门分公司无线 SSID 为 RUIJIE，用户接入无线局域网络时，采用 WPA2 加密方式，口令为 123456789。

在无线交换机上配置 OSPF 路由协议，实现和出口路由器的网络连通。

2. Windows 系统服务安装与配置

在总部的网络中，使用 4 台计算机搭建网络服务器群，便于全公司共享总部信息资源。其中，每台计算机使用 VirtualBox 安装两台虚拟计算机；接入层使用 1 台计算机，具体要求如下（注意：设置所有服务器口令为 ZXBzxb123#）。

（1）配置域控制器

① 在计算机 1（PC1）上，使用 VirtualBox 安装 Windows Server 2008 R2 网络操作系统，分配内存为 1GB，硬盘为 40GB，合法域名为 dc.zxb.com，IP 地址为 10.0.5.8/24。

② 在 VirtualBox 上添加两块虚拟硬盘，分配每块硬盘的大小为 20GB。将两块硬盘制作成 RAID1，盘符为 E:\。

③ 配置计算机 1（PC1）上搭建的服务器为域控制器，合法域名为 zxb.com。创建 4 个组，组名采用部门名称拼音，每个部门中创建 5 个用户。其中，销售部用户为 user1～user5；营销部用户为 user6～user10；市场部用户为 user11～user15；管理部用户为 user16～user20。用户名采用"user+数字"方式，如"user1"用户不能修改用户口令，用户口令为 ZXBzxb123#。密码最长使用期限为 9 天，最小长度为 9，锁定阈值为 3 次。要求用户只能在上班时间（每

天 9:00～18:00）登录。

④ 将计算机上搭建的服务器配置为主 DNS 服务器，正确配置 zxb.com 域名正向区域与 IPv4 反向区域，能够正确解析网络中的所有服务器。创建所有服务器主机记录和邮件服务器 MX 记录，关闭网络掩码排序功能。当遇到无法解析的域名时，将请求转发至 201.106.0.20 互联网域名服务器，只允许服务器 10.0.5.9 进行复制。

设置 DNS 服务正向区域和反向区域与活动目录集成；要求动态更新设置为非安全。

⑤ 制订网络服务器备份计划，每天 0 点对系统状态备份，备份信息保存至 E:\盘。

（2）配置 FTP 服务器

① 在计算机 1（PC1）上，使用 VirtualBox 安装 Windows Server 2008 R2 操作系统，分配其内存为 1GB，硬盘为 40GB，合法域名为 ftp.zxb.com，IP 地址为 10.0.5.14/24、10.0.5.15/24、10.0.5.16/24，将服务器加入到 Windows 域中。

② 在 VirtualBox 上添加 3 块虚拟硬盘，每块硬盘大小为 2GB。将 3 块硬盘制作成 RAID5 格式，盘符为 E:\。安装 IIS7.0 中的组件 FTP 服务，创建 3 个 FTP 站点，分别为 ftp.zxb.com、ftp1.zxb.com、ftp2.zxb.com，分别对应 IP 地址 10.0.5.14、10.0.5.15、10.0.5.16。其根目录分别为 E:\ftp、E:\ftp1、E:\ftp2。当同一个用户登录不同站点时，其主目录分别为 E:\ftp、E:\ftp1、E:\ftp2。

③ 其中，ftp.zxb.com 站点只允许销售部用户上传和下载文件，其他部门用户及匿名用户只能下载文件，不能上传文件。此外，限制销售部中用户上传的最大文件为 100MB，超过 80MB 时要有预警。ftp1.zxb.com 站点只允许营销部上传和下载文件，其他部门用户及匿名用户只能下载文件，不能上传文件，限制营销部中用户上传的最大文件为 100MB，超过 80MB 时要有预警。

ftp2.zxb.com 站点只允许市场部和管理部上传和下载文件，其他部门用户及匿名用户只能下载文件，不能上传文件，限制这两个部门中用户上传的最大文件为 100MB，超过 80MB 时要有预警。

④ 设置 FTP 服务目录列表样式为 UNIX 系统，设置控制通道超时 100s，数据通道超时 20s，最大连接数为 1000，FTP 拒绝扩展名为 .exe、.vbs 的文件请求对。

⑤ 在 VirtualBox 上添加 3 块虚拟硬盘，每块硬盘的大小为 5GB。将 3 块硬盘制作成 RAID5 格式，磁盘盘符为 F:\。为了防止误删除 FTP 站点中的文件，需要在 E:\盘中启用卷影副本功能，设置每周一至周五中午 12:30 进行卷影副本，将副本存在 F:\盘中。

（3）配置 DHCP 服务器

① 在计算机 2（PC2）上，使用 VirtualBox 安装 Windows Server 2008 R2 Core 操作系统，分配其内存为 1GB，硬盘为 40GB，合法域名为 dhcp.zxb.comIP，相应 IP 地址为 10.0.5.11/24。

② 将该服务器加入 Windows 域环境。

③ 在该服务器上配置 DHCP 服务，为内网 VLAN 10、VLAN 20、VLAN 30 和 VLAN 40 中的用户主机动态分配 IP 地址。建立作用域，作用域的名称为相应 VLAN 的名称，超级作用域名称为部门名称全拼。为内网中用户分配网关、DNS 服务器及域名。其中，VLAN 10 地址段为 10.0.1.10～10.0.1.200；VLAN 20 地址段为 10.0.2.10～10.0.2.200；VLAN 30 地址段为 10.0.3.10～10.0.3.200；VLAN 40 地址段为 10.0.4.10～10.0.4.200。

（4）配置客户机

① 在计算机 4（PC4）上，使用 VirtualBox 安装 Windows 7 操作系统，分配内存为 512MB，

硬盘为 40GB；将计算机加入域中，合法域名为 pc1.zxb.com，IP 地址为动态方式获取。

② 在计算机上安装无线网卡，测试是否正常登录到无线网络中。

（5）配置客户机

① 在计算机 4（PC4）上，使用 VirtualBox 安装 Windows 7 单机操作系统，分配其内存为 512MB，硬盘为 40GB。将计算机加入域中，合法域名为 pc2.zxb.com，IP 地址为动态方式获取。

② 检测该计算机是否能正常获取 IP 地址。

3. Linux 系统服务安装与配置

（1）安装 Web 服务器

① 在计算机 2（PC2）上，使用 VirtualBox 安装 CentOS 6.5 操作系统，分配其内存为 512MB，硬盘为 40GB。将服务器加入 Windows 域中，合法域名为 www.zxb.com，网络服务 IP 地址为 10.0.5.12/24。

② 安装 Apache 组件，在/var/www/目录下创建 zxb.com 文件夹，在文件夹中创建名称为 zxb.html 的主页文件，主页内容为"热烈庆祝 XXX 职业技能竞赛开幕"；使用虚拟主机创建并发布 www.zxb.com 站点；使用 openssl 申请证书，要求只允许使用域名通过 SSL 加密访问。

③ 设置网站响应时间及超时时间为 100s，服务器连接提出请求数量为 100，服务器两次请求时间间隔为 10s，服务器启动时运行进程数为 12，服务器最大空闲进程数为 25，服务器允许客户端最大并发连接数 30，服务器子进程所请求数量为 4500。设置 Apache 进程运行的用户和组为 nobody。

④ 在服务器上使用 iptables 设置防火墙功能，只允许用户访问这台服务器的 WWW 服务，服务器只能被动地接收连接请求，不能主动发起连接。允许服务器与 NFS 服务器建立 NFS 服务连接。

⑤ 将/var/www/zxb.com 主目录挂载到 NFS 服务器所提供的网络存储空间中，并要求开机自动挂载。设置 WWW 服务为级别 3 和 5 时开机启动。

（2）安装备份 DNS 服务器

① 在计算机 3（PC3）上，使用 VirtualBox 安装 CentOS 6.5 操作系统，分配其合法域名为 bdns.zxb.com，网络服务 IP 地址为 10.0.5.9/24。

② 将所有主 DNS 区域都复制到备份 DNS 服务器上；在 VirtualBox 上添加一块虚拟硬盘，硬盘的大小为 5GB。对该磁盘进行分区和格式化，将该磁盘设置为主分区，格式化为 EXT3 格式，文件块的大小为 2048B。创建/zxb.com 目录，将该磁盘分区挂载至文件夹，并要求开机自动挂载。

③ 安装 NFS 服务，目录/zxb.com 为 WWW 服务提供存储空间，只允许 10.0.5.12 主机挂载共享目录，具备读写权限，同步写入内存与磁盘，用户映射的用户 ID 为 510，组 ID 为 510。

④ 在服务器上使用 iptables 设置防火墙功能，只允许用户访问这台服务器的 DNS 服务、DNS 区域传送及为 10.0.5.12 主机提供 NFS 服务。设置 NFS 服务、DNS 服务需要在运行级别为 3 和 5 时开机自动启动。

（3）安装 Mail 服务器

① 在计算机 3（PC3）上，使用 VirtualBox 安装 CentOS 6.5 操作系统，分配其内存为 512MB，硬盘为 40GB，合法域名为 mail.zxb.com，网络服务 IP 地址为 10.0.5.10/24。

多层交换技术（实践篇）

② 安装 SendMail 软件，为所有部门的用户创建邮箱，邮箱存储在数据中心提供的网络存储上；限制每个用户邮箱的最大存储空间为 50MB，当用户邮箱存储达到 40MB 时发出警告；设置 SMTP 连接数为 200，连接超时为 10min。

③ 在服务器上使用 iptables 设置防火墙，只允许用户访问这台服务器的 Mail 服务。

④ 设置 Mail 服务需要在运行级别为 3 和 5 时开机自动启动。

【实施步骤】

（1）按照图 36-1 所示网络拓扑给所有的设备命名。在接入网络中的所有网络设备接口、子接口、VLAN 接口上，按照要求配置 IP 地址信息。

```
Router#configure terminal
Router(config)#hostname Router1
Router1(config)#interface FastEthernet 0/1
Router1(config-if-FastEthernet 0/1)#ip address 10.0.1.1 255.255.255.0
Router1(config-if-FastEthernet 0/1)#no shutdown
Router1(config-if-FastEthernet 0/1)#exit

Router1(config)#interface Serial 1/0
Router1(config-if-Serial 1/0)#ip address 10.0.2.1 255.255.255.0
Router1(config-if-Serial 1/0)#no shutdown
Router1(config-if-Serial 1/0)#exit

Router1(config)#interface Serial 2/0
Router1(config-if-Serial 2/0)#ip address 10.0.3.1 255.255.255.0
Router1(config-if-Serial 2/0)#no shutdown
Router1(config-if-Serial 2/0)#exit
```

```
Router#configure terminal
Router(config)#hostname Router2
Router2(config)#interface FastEthernet 0/1
Router2(config-if-FastEthernet 0/1)#ip address 172.16.1.1  255.255.255.0
Router2(config-if-FastEthernet 0/1)#no shutdown
Router2(config-if-FastEthernet 0/1)#exit

Router2(config)#interface Serial 1/0
Router2(config-if-Serial 1/0)#ip address 10.0.2.2 255.255.255.0
Router2(config-if-Serial 1/0)#no shutdown
Router2(config-if-Serial 1/0)#exit
```

```
Router#configure terminal
Router(config)#hostname Router3
```

```
Router3(config)#interface Serial 2/0
Router3(config-if-Serial 2/0)#ip address 10.0.3.2 255.255.255.252
Router3(config-if-Serial 2/0)#no shutdown
Router3(config-if-Serial 2/0)#exit

Router3(config)#interface FastEthernet 0/1
Router3(config-if-FastEthernet 0/1)#ip address 10.0.4.1  255.255.255.252
Router3(config-if-FastEthernet 0/1)#no shutdown
Router3(config-if-FastEthernet 0/1)#exit


Switch# configure terminal
Switch(config)#hostname S3760E-1
S3760E-1(config)#interface Fastethernet 0/1
S3760E-1(config-if-FastEthernet 0/1)#no switch
S3760E-1(config-if-FastEthernet 0/1)#ip address 10.0.1.2 255.255.255.252
S3760E-1(config-if-FastEthernet 0/1)#no shutdown
S3760E-1(config-if-FastEthernet 0/1)#exit


S3760E-1(config)#vlan 10
S3760E-1(config-vlan)#vlan 20
S3760E-1(config-vlan)#vlan 30
S3760E-1(config-vlan)#vlan 40
S3760E-1(config-vlan)#vlan 50
S3760E-1(config-vlan)#exit


S3760E-1(config)#interface vlan 10
S3760E-1(config-if-vlan 10)#ip address 10.0.10.1 255.255.255.0
S3760E-1(config-if-vlan 10)#no shutdown
S3760E-1(config-if-vlan 10)#exit


S3760E-1(config)#interface vlan 20
S3760E-1(config-if-vlan 20)#ip address 10.0.20.1 255.255.255.0
S3760E-1(config-if-vlan 20)#no shutdown
S3760E-1(config-if-vlan 20)#exit


S3760E-1(config)#interface vlan 30
S3760E-1(config-if-vlan 30)#ip address 10.0.30.1 255.255.255.0
S3760E-1(config-if-vlan 30)#no shutdown
S3760E-1(config-if-vlan 30)#exit
```

多层交换技术（实践篇）

```
S3760E-1(config)#interface vlan 40
S3760E-1(config-if-vlan 40)#ip address 10.0.40.1 255.255.255.0
S3760E-1(config-if-vlan 40)#no shutdown
S3760E-1(config-if-vlan 40)#exit

S3760E-1(config)#interface vlan 50
S3760E-1(config-if-vlan 50)#ip address 10.0.5.1 255.255.255.0
S3760E-1(config-if-vlan 50)#no shutdown
S3760E-1(config-if-vlan 50)#exit

S3760E-1(config)#int range FastEthernet 0/23-24
S3760E-1(config-if-range)#switch mode trunk
S3760E-1(config-if-range)#no shutdown
S3760E-1(config-if-range)#exit
```

```
Switch# configure terminal
Switch(config)#hostname S3760E-2
S3760E-2(config)#int range FastEthernet 0/23-24
S3760E-2(config-if-range)#switch mode trunk
S3760E-2(config-if-range)#no shutdown
S3760E-2(config-if-range)#exit
```

（2）在所有交换机上创建相应 VLAN，每个 VLAN 使用部门名称拼音命名，如销售部 VLAN 的名称为"xiao-shou-bu"。

```
S3760E-1#configure terminal
S3760E-1(config)#vlan 10
S3760E-1(config-if-vlan 10)#name xiao-shou-bu
S3760E-1(config-if-vlan 10)#vlan 20
S3760E-1(config-if-vlan 20)#name ying-xiao-bu
S3760E-1(config-if-vlan 20)#vlan 30
S3760E-1(config-if-vlan 30)#name shi-chang-bu
S3760E-1(config-if-vlan 30)#vlan 40
S3760E-1(config-if-vlan 40)#name guan-li-bu
S3760E-1(config-if-vlan 40)#vlan 50
S3760E-1(config-if-vlan 50)#name fu-wu-qi-qun
S3760E-1(config-if-vlan 50)#exit
```

```
Switch#configure terminal
Switch(config)#hostname S3760E-2
S3760E-2(config)#vlan 10
S3760E-2(config-vlan)#name xiao-shou-bu
```

```
S3760E-2(config-vlan)#vlan 20
S3760E-2(config-vlan)#name ying-xiao-bu
S3760E-2(config-vlan)#vlan 30
S3760E-2(config-vlan)#name shi-chang-bu
S3760E-2(config-vlan)#vlan 40
S3760E-2(config-vlan)#name guan-li-bu
S3760E-2(config-vlan)#exit
```

（3）将交换机的相关接口划分到相应的 VLAN 中。

```
S3760E-1(config)#
S3760E-1(config)#int range FastEthernet 0/2-6
S3760E-1(config-if-range)#switch access vlan 50
S3760E-1(config-if-range)#exit


S3760E-2(config)#
S3760E-2(config)#int range FastEthernet 0/1-5
S3760E-2(config-if-range)#switch access vlan 10
S3760E-2(config-if-range)#exit

S3760E-2(config)#int range FastEthernet 0/6-10
S3760E-2(config-if-range)#switch access vlan 20
S3760E-2(config-if-range)#exit

S3760E-2(config)#int range FastEthernet 0/11-15
S3760E-2(config-if-range)#switch access vlan 30
S3760E-2(config-if-range)#exit

S3760E-2(config)#int range FastEthernet 0/16-20
S3760E-2(config-if-range)#switch access vlan 40
S3760E-2(config-if-range)#exit
```

（4）对于总部路由器与分公司连接的接口描述，如 Router1 路由器 S2/0 接口与泉州分公司的 Router3 路由器 S2/0 接口连接，其接口描述为"Router1-TO-Router2-interface-S2/0"，其他以此类推。

```
Router1(config)#interface Serial 2/0
Router1(config-if-Serial 2/0)#description Router1-TO-Router3-interface-S2/0
Router1(config-if-Serial 2/0)#no shutdown
Router1(config-if-Serial 2/0)#exit

Router1(config)#interface Serial 1/0
Router1(config-if-Serial 1/0)#description Router1-TO-Router2-interface-s1/0
Router1(config-if-Serial 1/0)#no shutdown
```

```
Router1(config-if-Serial 1/0)#exit

Router1(config)#interface FastEthernet0/1
Router1(config-if-FastEthernet0/1)#description Router1-TO-Firewall-interface-fa0/1
Router1(config-if-FastEthernet0/1)#no shutdown
Router1(config-if-FastEthernet0/1)#exit
```

（5）总部三层核心交换机与核心中的三层交换机之间使用双链路，采用 MSTP 创建两个生成树实例，分别为实例 10 和实例 20。将 VLAN 10 和 VLAN 20 加入实例 10，将 VLAN 30 和 VLAN 40 加入实例 20。VALN 10、VLAN 20 中的数据流量通过第一条链路传输；VALN 30、VLAN 40 中的数据流量通过第二条链路传输。

```
S3760E-1(config)#spanning-tree
S3760E-1(config)#spanning-tree mst configuration
S3760E-1(config-mst)#instance 10 vlan 10,20
S3760E-1(config-mst)#instance 20 vlan 30,40
S3760E-1(config-mst)#name test
S3760E-1(config-mst)#revision 1
S3760E-1(config-mst)#exit

S3760E-1(config)#spanning-tree mst 10 priority 4096
S3760E-1(config)#spanning-tree mst 20 priority 8192


S3760E-2(config)#spanning-tree
S3760E-2(config)#spanning-tree mst configuration
S3760E-2(config-mst)#instance 10 vlan 10,20
S3760E-2(config-mst)#instance 20 vlan 30,40
S3760E-2(config-mst)#name test
S3760E-2(config-mst)#revision 1
S3760E-2(config-mst)#exit

S3760E-2(config)#spanning-tree mst 20 priority 4096
S3760E-2(config)#spanning-tree mst 10 priority 8192
```

（6）在总部的网络中，在核心网络中的三层交换机上，部署 DHCP 监听技术。在总部核心网络的三层交换机上配置 DHCP 技术。

```
S3760E-1(config)#service dhcp
S3760E-1(config)#ip dhcp Relay Information Option

S3760E-1(config)#ip dhcp pool vlan 10
S3760E-1(dhcp-config)#network 10.0.10.0 255.255.255.0
S3760E-1(dhcp-config)#default-router 10.0.10.1
```

```
S3760E-1(dhcp-config)#exit

S3760E-1(config)#ip dhcp pool vlan 20
S3760E-1(dhcp-config)#network 10.0.20.0 255.255.255.0
S3760E-1(dhcp-config)#default-router 10.0.20.1
S3760E-1(dhcp-config)#exit

S3760E-1(config)#ip dhcp pool vlan 30
S3760E-1(dhcp-config)#network 10.0.30.0 255.255.255.0
S3760E-1(dhcp-config)#default-router 10.0.30.1
S3760E-1(dhcp-config)#exit

S3760E-1(config)#ip dhcp pool vlan 40
S3760E-1(dhcp-config)#network 10.0.40.0 255.255.255.0
S3760E-1(dhcp-config)#default-router 10.0.40.1
S3760E-1(dhcp-config)#exit

S3760E-1(config)#ip dhcp pool vlan 50
S3760E-1(dhcp-config)#network 10.0.50.0 255.255.255.0
S3760E-1(dhcp-config)#default-router 10.0.50.1
S3760E-1(dhcp-config)#exit
```

```
S3760E-2(config)#service dhcp
S3760E-2(config)#int vlan 10
S3760E-2(config-if-vlan 10)#ip helper-address 10.0.10.1
S3760E-2(config-if-vlan 10)#exit

S3760E-2(config)#int vlan 20
S3760E-2(config-if-vlan 20)#ip helper-address 10.0.20.1
S3760E-2(config-if-vlan 20)#exit

S3760E-2(config)#int vlan 30
S3760E-2(config-if-vlan 30)#ip helper-address 10.0.30.1
S3760E-2(config-if-vlan 30)#exit

S3760E-2(config)#int vlan 40
S3760E-2(config-if-vlan 40)#ip helper-address 10.0.40.1
S3760E-2(config-if-vlan 40)#exit
```

```
S3760E-2(config)#int vlan 50
S3760E-2(config-if-vlan 50)#ip helper-address 10.0.5.1
S3760E-2(config-if-vlan 50)#exit
```

（7）总部与分公司之间租用链路传输业务数据，在总部和两个分公司的接入路由器上使用 IPSec 技术对数据进行加密。VPN 需要采用隧道模式，预共享密码为 123456，数据采用 ESP-DES、ESP-HASH-MD5 加密方式，AH 采用 Hash 方式进行加密。

```
Router1(config)#ip route 0.0.0.0  0.0.0.0  10.0.2.2
Router1(config)#crypto isakmap policy 1
Router1(config-isakmap)#hash md5
Router1(config-isakmap)#encryption 3des
Router1(config-isakmap)#authentication pre-share

Router1(config)#crypto isakmap key 123456  address 10.0.2.2
Router1(config)#crypto ipsec transform-set vpn1 ah-md5-hmac esp-des
Router1(config)#access-list 101 permit ip 10.0.1.0 0.0.0.255 10.0.2.1 0.0.0.255

Router1(config)#crypto map vpn1-map 1 ipsec-isakmap
Router1(config-crypto-map)#set peer 10.0.2.2
Router1(config-crypto-map)#set transform-set vpn1
Router1(config-crypto-map)#match address 101

Router1(config)#interface Serial 1/0
Router1(config-if-Serial 1/0)#crypto map vpn1-map
Router1(config-if-Serial 1/0)#exit


Router2(config)#ip route 0.0.0.0  0.0.0.0  10.0. 2.1
Router2(config)#crypto isakmap policy 1
Router2(config-isakmap)#hash md5
Router2(config-isakmap)#encryption 3des
Router2(config-isakmap)#authentication pre-share

Router2(config)#crypto isakmap key 123456  address 10.0.2.1
Router2(config)#crypto ipsec transform-set  vpn1 ah-md5-hmac esp-des
Router2(config)#access-list 101 permit ip 10.0.2.1 0.0.0.255  10.0.1.0 0.0.0.255

Router2(config)#crypto map vpn1-map 1 ipsec-isakmap
Router2(config-crypto-map)#set peer 10.0.2.1
```

任务 36 综合实训 1

```
Router2(config-crypto-map)#set transform-set  vpn1
Router2(config-crypto-map)#match address 101

Router2(config)#interface Serial 1/0
Router2(config-if-serial 1/0)#crypto map vpn1-map
Router2(config-if-serial 1/0)#exit


Router1(config)#ip route 0.0.0.0  0.0.0.0  10.0.3.2
Router1(config)#crypto isakmap policy 2
Router1(config-isakmap)#hash md5
Router1(config-isakmap)#encryption 3des
Router1(config-isakmap)#authentication pre-share

Router1(config)#crypto isakmap key 123456  address 10.0.3.2
Router1(config)#crypto ipsec transform-set  vpn2 ah-md5-hmac esp-des
Router1(config)#access-list 102  permit ip 10.0.1.0 0.0.0.255 10.0.4.0 0.0.0.255

Router1(config)#crypto map vpn2-map 2 ipsec-isakmap
Router1(config-crypto-map)#set peer 10.0.3.2
Router1(config-crypto-map)#set transform-set vpn2
Router1(config-crypto-map)#match address 102

Router1(config)#interface serial 2/0
Router1(config-if-serial 2/0)# crypto map vpn2-map
Router1(config-if-serial 2/0)#exit


Router3(config)#ip route 0.0.0.0  0.0.0.0  10.0.3.1
Router3(config)#crypto isakmap policy 2
Router3(config-isakmap)#hash md5
Router3(config-isakmap)#encryption 3des
Router3(config-isakmap)#authentication pre-share

Router3(config)#crypto isakmap key 123456  address 10.0.3.1
Router3(config)#crypto ipsec transform-set  vpn2 ah-md5-hmac esp-des
Router3(config)#access-list 101 permit ip 10.0.4.0 0.0.0.255 10.0.1.0 0.0.0.255

Router3(config)#crypto map vpn2-map 1 ipsec-isakmap
```

多层交换技术（实践篇）

```
Router3(config-crypto-map)#set peer 10.0.3.1
Router3(config-crypto-map)#set transform-set vpn2
Router3(config-crypto-map)#match address 102

Router3(config)#interface serial 2/0
Router3(config-if-serial 2/0)# crypto map vpn2-map
Router3(config-if-serial 2/0)#exit
```

（8）在链路上使用 PPP 协议，并采用 CHAP 方式，以总公司为验证端、分公司为被验证端，口令为 ruijie。

```
Router1(config)#username Router3 password ruijie
Router1(config)#interface serial 2/0
Router1(config-if-Serial 2/0)#encapsulation ppp
Router1(config-if-Serial 2/0)#ppp authentication chap
Router1(config-if-Serial 2/0)#exit


Router3(config)#username Router1 password ruijie
Router3(config)#interface serial 2/0
Router3(config-if-Serial 2/0)#encapsulation ppp
Router3(config-if-Serial 2/0)#exit
```

（9）全网配置 OSPF 动态路由协议实现连通。其中，总公司网络设备 Router1 的 RID 为 1.1.1.1；Router2 的 RID 为 2.2.2.2，Router3 的 RID 为 3.3.3.3；S3760E-1 和 S3760E-2 的 RID 为 4.4.4.4。

```
Router1(config)#router ospf
Router1(config-router)#router-id 1.1.1.1
Router1(config-router)#network 10.0.1.0 0.0.0.255 area 0
Router1(config-router)#network 10.0.2.0 0.0.0.255 area 0
Router1(config-router)#network 10.0.3.0 0.0.0.255 area 0
Router1(config-router)#exit


Router2(config)#router ospf
Router2(config-router)#router-id 2.2.2.2
Router2(config-router)#network 10.0.2.0 0.0.0.255 area 0
Router2(config-router)#network 172.16.1.0 0.0.0.255 area 0
Router2(config-router)#exit


Router3(config)#router ospf
Router3(config-router)#router-id 3.3.3.3
Router3(config-router)#network 10.0.3.0 0.0.0.255 area 0
Router3(config-router)#network 10.0.4.0 0.0.0.255 area 0
```

```
Router3(config-router)#network 10.0.60.0 0.0.0.255 area 0
Router3(config-router)#network 10.0.70.0 0.0.0.255 area 0
Router3(config-router)#exit


S3760E-1(config)#router ospf
S3760E-1(config-router)#router-id 4.4.4.4
S3760E-1(config-router)#network 10.0.1.0 0.0.0.255 area 0
S3760E-1(config-router)#network 10.0.10.0 0.0.0.255 area 0
S3760E-1(config-router)#network 10.0.20.0 0.0.0.255 area 0
S3760E-1(config-router)#network 10.0.30.0 0.0.0.255 area 0
S3760E-1(config-router)#network 10.0.40.0 0.0.0.255 area 0
S3760E-1(config-router)#network 10.0.5.0 0.0.0.255 area 0
S3760E-1(config-router)#exit


S3760E-2(config)#router ospf
S3760E-2(config-router)#router-id 4.4.4.4
S3760E-2(config-router)#network 10.0.10.0 0.0.0.255 area 0
S3760E-2(config-router)#network 10.0.20.0 0.0.0.255 area 0
S3760E-2(config-router)#network 10.0.30.0 0.0.0.255 area 0
S3760E-2(config-router)#network 10.0.40.0 0.0.0.255 area 0
S3760E-2(config-router)#exit
```

（10）在 OSPF 动态路由协议的网络中，路由更新时采用基于接口的 MD5 验证方式，口令为 ruijie。

```
Router1(config)#interface serial 1/0
Router1(config-if-Serial 1/0)#ip ospf message-digest-key 1 md5 ruijie
Router1(config-if-Serial 1/0)#exit


Router1(config)#interface serial 2/0
Router1(config-if-Serial 2/0)#ip ospf message-digest-key 1 md5 ruijie
Router1(config-if-Serial 2/0)#exit


Router1(config)#router ospf
Router1(config-router)#area 0 authentication message-digest


Router2(config)#interface Serial 1/0
Router2(config-if-Serial 1/0)#ip ospf message-digest-key 1 md5 ruijie
Router2(config-if-Serial 1/0)#exit


Router3(config)#interface serial 2/0
```

多层交换技术（实践篇）

```
Router3(config-if-Serial 1/0)#ip ospf message-digest-key 1 md5 ruijie
Router3(config-if-Serial 1/0)#exit
```

（11）厦门分公司使用 NAT 技术，把内网中的私有地址转换为路由器的外部接口地址（全局地址），允许内网用户在上班时间（周一至周五 9:00~18:00）访问互联网。

```
Router2(config)#int FastEthernet 0/1
Router2(config-if-FastEthernet 0/1)#ip nat inside
Router2(config-if-FastEthernet 0/1)#exit

Router2(config)#int serial 1/0
Router2(config-if-Serial 1/0)#ip nat outside
Router2(config-if-Serial 1/0)#exit

Router2(config)#access-list 10 permit  any

Router2(config)#ip nat pool abc 10.0.2.2  10.0.2.2  netmask 255.255.255.0
Router2(config)#ip nat inside source list 10 pool abc overload

Router2(config)#time-range on-work
Router2(config time-range)#periodic weekdays 09:00 to 18:00
Router2(config time-range)#exit

Router2(config)#interface serial 1/0
Router2(config-if-Serial 1/0)#ip access-group 10 out
Router2(config-if-Serial 1/0)#no shutdown
Router2(config-if-Serial 1/0)#exit
```

（12）在厦门分公司路由器上配置默认路由，并使用路由重分发技术将默认路由发布到 OSPF 网络中，网络类型为 E1，开销为 220。

```
Router2(config)#
Router2(config)#ip route 0.0.0.0 0.0.0.0 s1/0

Router2(config)#router ospf
Router2(config-router)#default-metric 220
Router2(config-router)#redistribute connected subnets
Router2(config-router)#redistribute static subnets
Router2(config-router)#default-information originate
```

（13）泉州分公司租用专用链路与总公司相连，其内网采用二层交换机接入，内网有两个部门 VLAN，所以使用单臂路由实现 VLAN 间路由连通。

```
S2628G(config)#vlan 70
S2628G(config-vlan)#interface range FastEthernet 0/2-10
```

```
S2628G(config-if-range)#switch access vlan 70
S2628G(config-if-range)#no shutdown
S2628G(config-if-range)#exit

S2628G(config)#vlan 80
S2628G(config-vlan)#interface range FastEthernet 0/11-20
S2628G(config-if-range)#switch access vlan 80
S2628G(config-if-range)#no shutdown
S2628G(config-if-range)#exit

S2628G(config)#
S2628G(config)#interface FastEthernet 0/1
S2628G(config-if-FastEthernet 0/1)#switch mode trunk
S2628G(config-if-FastEthernet 0/1)#exit


Router3(config)#
Router3(config)#interface FastEthernet 0/1
Router3(config-if-FastEthernet 0/1)#no ip address
Router3(config-if-FastEthernet 0/1)#exit

Router3(config)#interface FastEthernet 0/0.70
Router3(config-subif-FastEthernet 0/0.70)#encapsulation dot1q 70
Router3(config-subif-FastEthernet 0/0.70)#ip address 10.0.70.1 255.255.255.0
Router3(config-subif-FastEthernet 0/0.70)#no shutdown
Router3(config-subif-FastEthernet 0/0.70)#exit

Router3(config)#interface FastEthernet 0/0.80
Router3(config-subif-FastEthernet 0/0.80)#encapsulation dot1q 80
Router3(config-subif-FastEthernet 0/0.80)#ip address 10.0.80.1 255.255.255.0
Router3(config-subif-FastEthernet 0/0.80)#no shutdown
Router3(config-subif-FastEthernet 0/0.80)#end
```

（14）在总部部署防火墙，配置防火墙为网桥模式，实现防火墙的高速传输。

登录防火墙设备，使用默认的账号"admin"和密码"admin"，进入防火墙的管理界面。选择"网络管理"→"接口"→"透明桥"选项，单击"新建"按钮，如图36-2所示。

填写透明桥的地址及其安全信息，其中，桥的 IP 地址为管理防火墙而设，不影响用户通信。勾选"ge1"和"ge2"复选框，表明将这两个接口划分到该桥中。勾选"HTTPS"

多层交换技术（实践篇）

"HTTP""TELNET""PING"等复选框，表示可以通过该桥对防火墙进行测试和管理，不影响用户通信，如图 36-3 所示。

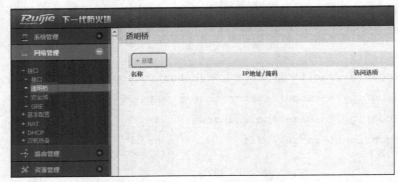

图 36-2　新建透明桥

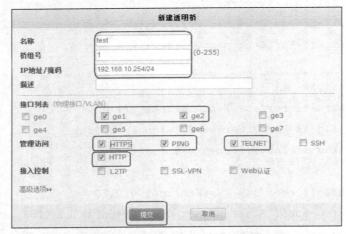

图 36-3　填写透明桥的地址及其安全信息

透明桥配置完成提交后的结果如图 36-4 所示。

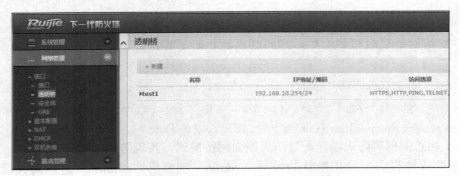

图 36-4　透明桥配置完成提交后的结果

（15）在总部的防火墙设备上配置防火墙策略，允许总部内网用户（VLAN 10、VLAN 20、VLAN 30、VLAN 40）每天 9:00～18:00 访问外部的互联网，允许网络服务器全天对外提供服务。

配置节点的地址范围（地址节点）。选择"资源管理"→"地址资源"→"地址节点"选项，单击"新建"按钮，如图 36-5 所示。

图 36-5　新建地址节点

填写名称和地址节点范围，并导入地址，单击"提交"按钮，如图 36-6 所示。

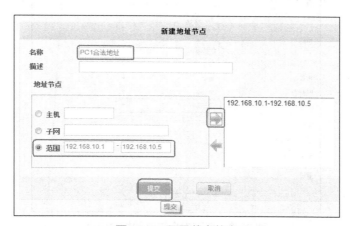

图 36-6　配置节点信息

节点配置完成提交后的结果如图 36-7 所示。

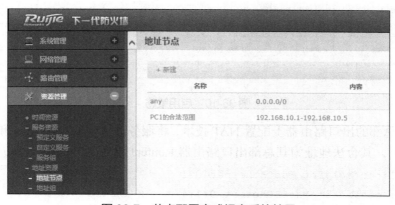

图 36-7　节点配置完成提交后的结果

多层交换技术（实践篇）

选择"防火墙"→"安全策略"→"安全策略"选项，单击"新建"按钮，如图 36-8 所示。

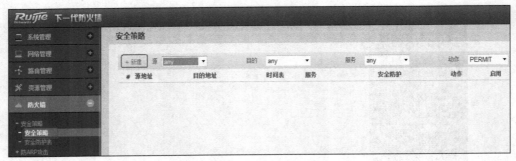

图 36-8　新建安全策略

填写安全策略的相关信息，如图 36-9 所示。

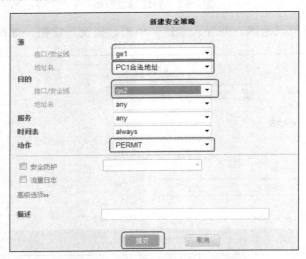

图 36-9　填写安全策略的相关信息

勾选"启用"复选框，如图 36-10 所示，启用刚才配置完成的"ge1->ge2 (0/1)"安全策略。

图 36-10　启用策略

（16）在总部的出口路由器上配置 NAT 技术，将服务器区中的 Web 服务和 FTP 服务发布到互联网上，其合法地址为其总部出口路由器 Router1 设备的外网接口 Fa0/2 的地址。

```
Router1(config)#int FastEthernet 0/1
Router1(config-if-FastEthernet 0/1)#ip nat inside
Router1(config-if-FastEthernet 0/1)#exit
```

```
Router1(config)#int serial 1/0
Router1(config-if-serial 1/0)#ip nat outside
Router1(config-if-serial 1/0)#exit

Router1(config)#ip nat inside source static tcp 10.0.5.12 80 10.0.2.1 80

Router1(config)#ip nat inside source static tcp 10.0.5.14 20 10.0.2.1 20
Router1(config)#ip nat inside source static tcp 10.0.5.14 21 10.0.2.1 21

Router1(config)#ip nat inside source static tcp 10.0.5.15 20 10.0.2.1 20
Router1(config)#ip nat inside source static tcp 10.0.5.15 21 10.0.2.1 21

Router1(config)#ip nat inside source static tcp 10.0.5.16 20 10.0.2.1 20
Router1(config)#ip nat inside source static tcp 10.0.5.16 21 10.0.2.1 21
```

（17）配置 "AC+Fit AP" 无线设备，满足厦门分公司内网使用无线覆盖的要求。

使用无线交换机作为 DHCP 服务器，为内部无线用户动态分配 IP 地址、网关和 DNS 服务器（218.85.152.99）；其分配地址段为 10.0.9.10～10.0.9.200。

配置 "AC+Fit AP" 结构的无线局域网环境，实现厦门分公司内网的无线连通，并能够访问外部互联网。

创建厦门分公司的无线 SSID 为 RUIJIE，用户接入无线局域网络时，需要采用 WPA2 加密方式，口令为 123456789。将连接 AP 的 AC 接口设置为全双工。

在无线交换机上配置 OSPF 动态路由协议，实现和出口路由器的连通。

① 配置 AP。

```
Ruijie#configure terminal
Ruijie(config)#ap-mode fit
```

② 配置 AC。

```
Ruijie(config)#
Ruijie(config)#hostname Ruijie-AC
Ruijie-AC(config)#interface GigabitEthernet 0/1
Ruijie-AC(config-if-GigabitEthernet 0/1)#ip address 172.16.1.2  255.255.255.0
Ruijie-AC(config-if-GigabitEthernet 0/1)#no shutdown
Ruijie-AC(config-if-GigabitEthernet 0/1)#exit

Ruijie-AC(config)#interface GigabitEthernet 0/0
Ruijie-AC(config-if-GigabitEthernet 0/0)#switchport mode trunk
Ruijie-AC(config-if-GigabitEthernet 0/0)#exit

Ruijie-AC(config)#vlan 90              // 无线用户 VLAN
Ruijie-AC(config-vlan)#exit
```

多层交换技术（实践篇）

```
Ruijie-AC(config)#interface vlan 90
Ruijie-AC(config-if-vlan 90)#ip address 10.0.90.1 255.255.255.0
Ruijie-AC(config-if-vlan 90)#no shutdown
Ruijie-AC(config-if-vlan 90)#exit

Ruijie-AC(config)#service dhcp
Ruijie-AC(dhcp-config)#ip dhcp pool VLAN 90
Ruijie-AC(dhcp-config)#option 138 ip 9.9.9.9
Ruijie-AC(dhcp-config)#network 10.0.90.0 255.255.255.0
Ruijie-AC(dhcp-config)#ip dhcp excluded-address 10.0.90.0 10.0.90.9
Ruijie-AC(dhcp-config)#ip dhcp excluded-address 10.0.90.201 10.0.90.255
Ruijie-AC(dhcp-config)#default-router 10.0.90.1
Ruijie-AC(dhcp-config)#exit

Ruijie-AC(config)#int vlan 90
Ruijie-AC(config-if-vlan 90)#ip helper-address 10.0.90.1
Ruijie-AC(config-if-vlan 90)#exit

Ruijie-AC(config)#interface GigabitEthernet 0/0
Ruijie-AC(config-if-GigabitEthernet 0/0)#duplex full
Ruijie-AC(config-if-GigabitEthernet 0/0)#no shutdown
Ruijie-AC(config-if-GigabitEthernet 0/0)#exit

Ruijie-AC(config)#vlan 80         // 无线设备 VLAN
Ruijie-AC(config-vlan)#exit
Ruijie-AC(config)#interface vlan 80
Ruijie-AC(config-if-vlan 80)#no shutdown
Ruijie-AC(config-if-vlan 80)#exit

Ruijie-AC(config)#interface loopback 0
Ruijie-AC(config-int-loopback 0)#ip address 9.9.9.9 255.255.255.255
// 必须是 Loopback 0，用于 AP 寻找 AC 的地址，为 DHCP 中的 option138 字段
Ruijie-AC(config-int-Loopback)#exit

Ruijie-AC(config)#interface GigabitEthernet 0/0     // 连接 AP 接口
Ruijie-AC(config-int-GigabitEthernet 0/0)#switchport access vlan 1
Ruijie-AC(config-int-GigabitEthernet 0/0)#exit

Ruijie(config)#wlan-config 1 ruijie
Ruijie(config-wlan)#exit
```

```
Ruijie(config)#ap-group abc
Ruijie(config-ap-group)#interface-mapping 1 90
Ruijie(config-ap-group)#exit

Ruijie(config)#ap-config 5869.6c84.1274
/* 把 AP 组的配置关联到 AP 上（XXX 为某个 AP 的名称时，表示只在该 AP 下应用 ap-group；第一次部署时，默认 XXX 实际上是 AP 的 MAC 地址）*/
Ruijie(config-ap-config)#ap-group abc
Ruijie(config-ap-group)#exit

Ruijie-AC(config)#wlansec 1
Ruijie-AC(config-wlansec)#security wpa enable
Ruijie-AC(config-wlansec)#security wpa ciphers aes enable
Ruijie-AC(config-wlansec)#security wpa akm psk enable
Ruijie-AC(config-wlansec)#security wpa akm psk set-key ascii 12345678
```

（18）在无线交换机上配置 OSPF 路由协议。

```
Ruijie-AC(config)#router ospf
Ruijie-AC(config-router)#network 172.16.1.0   0.0.0.255   area 0
Ruijie-AC(config-router)#network 10.0.80.0    0.0.0.255   area 0
Ruijie-AC(config-router)#network 10.0.90.0    0.0.0.255   area 0
Ruijie-AC(config-router)#exit
```

备注：关于网络服务器的配置，由于篇幅有限，此处省略。

任务 37 综合实训 2

【任务描述】

某大学本部校区在上海，分别在北京和广州建立了两个分校区，为保证上海本部与分校之间的网络连通，在上海本部校区出口规划了双出口及双链路备份，确保所有校区正常访问互联网。

保障上海本部与分校之间的网络连通的同时，需要对某些业务进行互访限制。另外，各校区的业务对网络可靠性要求较高，要求网络核心区域发生故障时，网络中断时间尽可能短。此外，网络部署时要考虑到网络的可管理性，并合理利用网络资源。

【组网拓扑】

图 37-1 所示网络拓扑为某大学上海本部和分校区的网络连接场景。其中，上海本部校区的两台出口网关 EG（编号分别为 EG1、EG2）与教育网和联通网互连，两台核心网络中的三层交换机（编号为 S3、S4）作为上海本部校区的核心网络中的交换机。其中，核心网络中的一台三层交换机（编号为 S5）实现服务器高速接入；上海本部校区的两台无线控制器 WS（编号为 AC1、AC2）用作无线接入点，实现上海本部的无线局域网接入；上海本部校区的两台二层接入交换机（编号为 S1、S2）用作上海本部校区的接入设备；上海本部校区的一台无线 AP（编号 AP1）用作上海本部校区的无线接入设备。

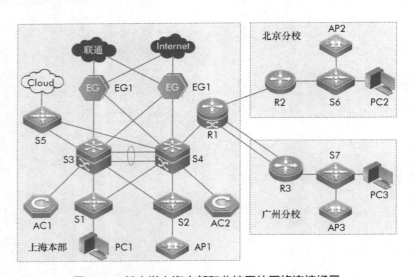

图 37-1 某大学上海本部和分校区的网络连接场景

任务 37 综合实训 2

3 个校区之间使用 3 台路由器 RSR20（编号为 R1、R2、R3）实现接入。其中，总部通过 R1 路由器与北京校区及广州校区的 R2 和 R3 路由器相连，两台核心网络中的三层交换机（编号为 S6、S7）分别作为北京分校和广州分校的核心网络中的交换机，两台无线 AP（编号 AP2、AP3）分别作为北京分校和广州分校的无线接入点。

详细的网络连接信息如表 37-1 所示。

表 37-1 详细的网络连接信息

设备名称	设备接口	接口描述	目标设备	设备接口
S1	Gi0/1	Con_To_PC1	PC1	
S1	Gi0/23	Con_To_S3_Gi0/1	S3	Gi0/1
S1	Gi0/24	Con_To_S4_Gi0/1	S4	Gi0/1
S2	Gi0/22	Con_To_AP1_Gi0/1	AP1	Gi0/1
S2	Gi0/23	Con_To_S3_Gi0/2	S3	Gi0/2
S2	Gi0/24	Con_To_S4_Gi0/2	S4	Gi0/2
S3	Gi0/1	Con_To_S1_Gi0/23	S1	Gi0/23
S3	Gi0/2	Con_To_S2_Gi0/23	S2	Gi0/23
S3	Gi0/3	Con_To_AC1_Gi0/1	AC1	Gi0/1
S3	Gi0/4	Con_To_EG2_Gi0/3	EG2	Gi0/3
S3	Gi0/5	Con_To_S5_Gi0/23	S5	Gi0/23
S3	Gi0/6	Con_To_EG1_Gi0/0	EG1	Gi0/0
S3	Te0/49	Con_To_S4_Te0/49	S4	Te0/49
S3	Te0/50	Con_To_S4_Te0/50	S4	Te0/50
S4	Gi0/1	Con_To_S1_Gi0/24	S1	Gi0/24
S4	Gi0/2	Con_To_S2_Gi0/24	S2	Gi0/24
S4	Gi0/3	Con_To_AC2_Gi0/1	AC2	Gi0/1
S4	Gi0/4	Con_To_EG1_Gi0/3	EG1	Gi0/3
S4	Gi0/5	Con_To_S5_Gi0/24	S5	Gi0/24
S4	Gi0/6	Con_To_EG2_Gi0/0	EG2	Gi0/0
S4	Gi0/7	Con_To_R1_Gi0/0	R1	Gi0/0
S4	Te0/49	Con_To_S3_Te0/49	S3	Te0/49
S4	Te0/50	Con_To_S3_Te0/50	S3	Te0/50
S5	Gi0/1	Con_To_Jcos	云服务器	
S5	Gi0/23	Con_To_S3_Gi0/5	S3	Gi0/5

多层交换技术（实践篇）

续表

设备名称	设备接口	接口描述	目标设备	设备接口
S5	Gi0/24	Con_To_S4_Gi0/5	S4	Gi0/5
AC1	Gi0/1	Con_To_S3_Gi0/3	S3	Gi0/3
AC2	Gi0/1	Con_To_S4_Gi0/3	S4	Gi0/3
EG1	Gi0/0	Con_To_S3_Gi0/6	S3	Gi0/6
EG1	Gi0/1	Con_To_ChinaUnicom	EG2	Gi0/1
EG1	Gi0/2	Con_To_Cernet	EG2	Gi0/2
EG1	Gi0/3	Con_To_S4_Gi0/4	S4	Gi0/4
EG2	Gi0/0	Con_To_S4_Gi0/6	S4	Gi0/6
EG2	Gi0/1	Con_To_ChinaUnicom	EG1	Gi0/1
EG2	Gi0/2	Con_To_Cernet	EG1	Gi0/2
EG2	Gi0/3	Con_To_S3_Gi0/4	S3	Gi0/4
R1	Gi0/0	Con_To_S4_Gi0/7	S4	Gi0/7
R1	S2/0	Con_To_R2_S2/0	R2	S2/0
R1	S3/0	Con_To_R3_S3/0	R3	S3/0
R1	S4/0	Con_To_R3_S4/0	R3	S4/0
R2	S2/0	Con_To_R1_S2/0	R1	S2/0
R2	Gi0/0	Con_To_S6_Gi0/24	S6	Gi0/24
R3	S3/0	Con_To_R1_S3/0	R1	S3/0
R3	S4/0	Con_To_R1_S4/0	R1	S4/0
R3	Gi0/0	Con_To_S7_Gi0/24	S7	Gi0/24
S6	Gi0/1	Con_To_PC2	PC2	
S6	Gi0/23	Con_To_AP2_Gi0/1	AP2	Gi0/1
S6	Gi0/24	Con_To_R2_Gi0/0	R2	Gi0/0
S7	Gi0/1	Con_To_PC3	PC3	
S7	Gi0/23	Con_To_AP3_Gi0/1	AP3	Gi0/1
S7	Gi0/24	Con_To_R3_Gi0/0	R3	Gi0/0

【设备清单】

本综合实训中所使用的网络设备及线缆如表 37-2 所示。

任务 37 综合实训 2

表 37-2 网络设备及线缆

序号	类别	设备	厂商	型号	数量
1	硬件	出口网关	锐捷	RG-EG 系列	2 台
2	硬件	路由器	锐捷	RG-RSR 系列	3 台
3	硬件	串口接口模块	锐捷	RG-SIC-1HS	6 个
4	硬件	串口线缆	锐捷	CAB-V.35DTE-V.35DCE	3 条
5	硬件	数据中心交换机	锐捷	RG-S6000C 系列	2 台
6	硬件	电源模块	锐捷	RG-PA70I	2 台
7	硬件	VSU 堆叠电缆	锐捷	XG-SFP-CU1M	2 条
8	硬件	三层交换机	锐捷	RG-S57 系列	3 台
9	硬件	二层接入交换机	锐捷	RG-S29 系列	2 台
10	硬件	无线控制器	锐捷	RG-WS 系列	2 台
11	硬件	无线 AP	锐捷	RG-AP 系列	3 台
12	硬件	电源适配器	锐捷	RG-E-120	3 个
13	软件	无线地勘系统	锐捷	锐捷无线地勘系统	1 套

【任务目标一】设备基础信息配置

1. 设备命名规则和设备的基础信息

（1）根据总体规划内容，对所有设备名称根据命名规则进行修订。

（2）依据设备的总体规划和详细的网络连接信息配置设备的接口描述信息。

2. 密码恢复和软件版本统一

（1）在三层交换机 S1 和 S2 上配置密码恢复，将新的密码设置为 ruijie。

（2）将三层交换机 S1 和 S2 的系统软件版本统一，更新版本至 RGOS 11.4(1)B1P3。

3. 网络设备安全技术

（1）在路由器和无线控制器上开启 SSH 服务器端功能，配置用户名和密码为 admin，密码为明文类型，特权密码为 admin。

（2）在所有交换机上开启 Telnet 功能，对所有 Telnet 用户采用本地认证方式。创建本地用户，配置用户名和密码为 admin，密码为明文类型，特权密码为 admin。

（3）在所有设备上配置 SNMP 消息，向主机 172.16.0.254 发送的 Trap 消息版本采用 V2C，读写的 Community 为 ruijie，只读的 Community 为 public，开启 Trap 消息。

【任务目标二】网络搭建与网络冗余备份方案部署

1. 虚拟局域网及 IPv4 地址部署

为了减少广播，需要规划并配置 VLAN，具体要求如下。

多层交换技术（实践篇）

（1）配置合理，Trunk 链路上不允许不必要的 VLAN 的数据流通过。

（2）为隔离网络中部分终端用户间的二层互访，在二层接入交换机 S1、S2 上开启端口保护功能。

根据上述要求及表 37-3、表 37-4，在各设备上完成 VLAN 配置和端口分配，并按照要求配置 IPv4 地址信息。

表 37-3　网络设备名称

拓扑图中设备名称	配置主机名（Hostname 名）
S1	BB-S2910-01
S2	BB-S2910-02
S3	BB-S6000-01
S4	BB-S6000-02
S5	BB-S5750-01
S6	BJFX-S5750-01
S7	GZFX-S5750-01
R1	BB-RSR20-01
R2	BJFX-RSR20-01
R3	GZFX-RSR20-01
AC1	BB-WS6008-01
AC2	BB-WS6008-02
EG1	BB-EG2000-01
EG2	BB-EG2000-02
AP1	BB-AP520-01
AP2	BJFX-AP520-01
AP3	GZFX-AP520-01

表 37-4　IPv4 地址分配信息

设备	接口或 VLAN	VLAN 名称	二层或三层规划（XX 现场提供）	说明
S1	VLAN 10	Office 10	Gi0/1～Gi0/4	办公网段
	VLAN 20	Office 20	Gi0/5～Gi0/8	办公网段
	VLAN 30	Office 30	Gi0/9～Gi0/12	办公网段
	VLAN 40	Office 40	Gi0/13～Gi0/16	办公网段
	VLAN 50	AP	Gi0/21～Gi0/22	无线 AP 管理
	VLAN 100	Manage	192.XX.100.4/24	设备管理 VLAN

续表

设备	接口或 VLAN	VLAN 名称	二层或三层规划（XX 现场提供）	说明
S2	VLAN 10	Office 10	Gi0/1～Gi0/4	办公网段
	VLAN 20	Office 20	Gi0/5～Gi0/8	办公网段
	VLAN 30	Office 30	Gi0/9～Gi0/12	办公网段
	VLAN 40	Office 40	Gi0/13～Gi0/16	办公网段
	VLAN 50	AP	Gi0/21～Gi0/22	无线 AP 管理
	VLAN 100	Manage	192.XX.100.5/24	设备管理 VLAN
S3	VLAN 10	Office 10	192.XX.10.252/24	办公网段
	VLAN 20	Office 20	192.XX.20.252/24	办公网段
	VLAN 30	Office 30	192.XX.30.252/24	办公网段
	VLAN 40	Office 40	192.XX.40.252/24	办公网段
	VLAN 50	AP	192.XX.50.252/24	无线 AP 管理
	VLAN 100	Manage	192.XX.100.252/24	设备管理 VLAN
	Gi0/1	Trunk	#	#
	Gi0/2	Trunk	#	#
	Gi0/3	Trunk	#	#
	Gi0/4	#	10.XX.0.41/30	#
	Gi0/5	#	10.XX.0.1/30	#
	Gi0/6	#	10.XX.0.5/30	互连 EG1
	LoopBack 0	#	11.XX.0.33/32	#
S4	VLAN 10	Office10	192.XX.10.253/24	办公网段
	VLAN 20	Office20	192.XX.20.253/24	办公网段
	VLAN 30	Office30	192.XX.30.253/24	办公网段
	VLAN 40	Office40	192.XX.40.253/24	办公网段
	VLAN 50	AP	192.XX.50.253/24	无线 AP 管理
	VLAN 100	Manage	192.XX.100.253/24	设备管理 VLAN
	Gi0/1	Trunk	#	#
	Gi0/2	Trunk	#	#
	Gi0/3	Trunk	#	#
	Gi0/4	#	10.XX.0.37/30	#
	Gi0/5	#	10.XX.0.33/30	#

多层交换技术（实践篇）

续表

设备	接口或 VLAN	VLAN 名称	二层或三层规划（XX 现场提供）	说明
S4	Gi0/6	#	10.XX.0.9/30	互连 EG2
	Gi0/7	#	10.XX.0.13/30	#
	Loopback 0	#	11.XX.0.34/32	#
AC1	Loopback 0	#	11.XX.0.204/32	#
	VLAN 60	Wireless	192.XX.60.252/24	无线用户
	VLAN 100	Manage	192.XX.100.2/24	管理与互连 VLAN
AC2	Loopback 0	#	11.XX.0.205/32	#
	VLAN60	Wireless	192.XX.60.253/24	无线用户
	VLAN100	Manage	192.XX.100.3/24	管理与互连 VLAN
S5	Gi0/1	#	193.XX.0.1/24	互连云平台
	Loopback 0	#	11.XX.0.5/32	#
	Gi0/23	#	10.XX.0.2/30	#
	Gi0/24	#	10.XX.0.34/30	#
EG1	Gi0/0	#	10.XX.0.6/30	#
	Gi0/1	#	196.XX.0.1/24	与 EG2 互连
	Gi0/2	#	197.XX.0.1/24	与 EG2 互连
	Gi0/3	#	10.XX.0.38/30	#
	Loopback 0	#	11.XX.0.11/32	#
EG2	Gi0/0	#	10.XX.0.10/30	#
	Gi0/1	#	196.XX.0.2/24	与 EG1 互连
	Gi0/2	#	197.XX.0.2/24	与 EG1 互连
	Gi0/3	#	10.XX.0.42/30	#
	Loopback 0	#	11.XX.0.12/32	#
R1	Gi0/0	#	10.XX.0.14/30	#
	S2/0	#	10.XX.0.18/30	#
	S3/0	#	10.XX.0.22/30	捆绑组 1 成员
	S4/0	#	10.XX.0.22/30	捆绑组 1 成员
	Loopback 0	#	11.XX.0.1/32	#
R2	Gi0/0	#	10.XX.0.25/30	#
	S2/0	#	10.XX.0.17/30	#
	Loopback 0	#	11.XX.0.2/32	#

续表

设备	接口或 VLAN	VLAN 名称	二层或三层规划（XX 现场提供）	说明
R3	Gi0/0	#	10.XX.0.29/30	#
	S3/0	#	10.XX.0.21/30	捆绑组 1 成员
	S4/0	#	10.XX.0.21/30	捆绑组 1 成员
	Loopback 0	#	11.XX.0.3/32	#
S6	Gi0/24	#	10.XX.0.26/30	#
	VLAN10	Wire_user	194.XX.10.254/24	分校有线用户
	VLAN20	Wireless_user	194.XX.20.254/24	分校无线用户
	VLAN30	AP	194.XX.30.254/24	分校无线 AP
	Loopback 0	#	11.XX.0.6/32	#
S7	Gi0/24	#	10.XX.0.30/30	#
	VLAN10	Wire_user	195.XX.10.254/24	分校有线用户
	VLAN20	Wireless_user	195.XX.20.254/24	分校无线用户
	VLAN30	AP	195.XX.30.254/24	分校无线 AP
	Loopback 0	#	11.XX.0.7/32	#
测试计算机	PC1	#	自动获取	#
	PC2	#	194.XX.10.2/24	#
	PC3	#	195.XX.10.2/24	#

2. 局域网环路规避方案部署

为了避免在网络的末端接入设备上出现环路而影响全网，要求在上海本部校区网络中的二层接入交换机设备 S1、S2 与分校网络核心中的三层交换机 S6、S7 上进行防环处理，具体要求如下。

（1）二层接入交换机接入接口开启 BPDU 防护，不能接收 BPDU Guard 报文。

（2）二层接入交换机接口下开启 RLDP 功能防止环路，检测到环路后处理方式为 shutdown-port。

（3）二层接入交换机连接终端的所有端口配置为边缘端口。

（4）如果二层接入交换机端口被 BPDU Guard 检测进入 Error-disabled 状态，再过 300s 后自动恢复，应重新检测是否有环路。

3. 接入安全部署

为了保证接入设备的 DHCP 服务安全，防止伪源 IP 地址攻击及 DOS 攻击，需要实施以下 DHCP 安全配置，具体要求如下。

（1）在上海本部校区的网络核心中的三层交换机 S3 上搭建 DHCP 服务器，实现对 VLAN 10 以内的用户 IP 地址的分配。

（2）为了防止从非法 DHCP 服务器获得地址而产生冲突，要求在上海本部校区的二层接入交换机 S1、S2 的上连接口的出方向使用 ACL（编号 101），限制非法 DHCP 服务器接入。

（3）为了防止大量网关发送的正常的相关报文被二层接入交换机误认为是攻击而丢弃，导致下连用户因为无法获取网关的 ARP 信息而无法上网，要求关闭上海本部校区的二层接入交换机 S1、S2 的上接口的 NFPP 功能。

（4）开启上海本部校区的核心网络中的三层交换机 S5 上的防 LAND 攻击功能，防非法 TCP 报文攻击，防自身消耗攻击，规避服务器 DOS 攻击。

4．MSTP 及 VRRP 部署

在上海本部校区的核心网络中的三层交换机 S3、S4 上配置 MSTP，防止二层环路。要求 VLAN 10、VLAN 20、VLAN 30 的数据流经过核心网络中的三层交换机 S3 转发；VLAN 40、VLAN 50、VLAN 100 的数据流经过核心网络中的三层交换机 S4 转发。此外，核心网络中的三层交换机 S3、S4 中任何一台宕机时，均可无缝切换至另一台进行转发。

所配置的参数要求如下。

（1）配置主域名称为 ruijie。
（2）配置版本为 1。
（3）配置实例 1，包含 VLAN 10、VLAN 20、VLAN 30。
（4）配置实例 2，包含 VLAN 40、VLAN50、VLAN100。
（5）配置 S3 作为实例 0、1 中的主根，S4 作为实例 0、1 的备份根。
（6）配置 S4 作为实例 2 中的主根，S3 作为实例 2 的备份根。
（7）配置主根优先级为 4096，备份根优先级为 8192。
（8）在 S3 和 S4 上配置 VRRP，实现主机的网关冗余。
（9）将 S3、S4 上各 VRRP 组中的高优先级设置为 150，低优先级设置为 120。

所配置的核心网络中的三层交换机 S3 和 S4 的 VRRP 参数要求如表 37-5 所示。

表 37-5　核心网络中的三层交换机 S3 和 S4 的 VRRP 参数要求

VLAN	VRRP 备份组号（VRID）	VRRP 虚拟 IP 地址
VLAN 10	10	192.XX.10.254
VLAN 20	20	192.XX.20.254
VLAN 30	30	192.XX.30.254
VLAN 40	40	192.XX.40.254
VLAN 50	50	192.XX.50.254
VLAN 100（交换机间）	100	192.XX.100.254

5．路由协议部署

上海本部内网中使用静态路由、OSPF 多协议组网，其中，上海本部内网中核心网络中的交换机 S3、S4、S5 的出口网关 EG1、EG2 和出口路由器 R1 上使用 OSPF 协议实现互连，其余三层设备之间使用静态路由协议互连。上海本部与两个分校广域网之间使用静态路由协议互连，各分校局域网内部网络使用 RIP 路由协议互连。

路由协议部署具体要求如下。
（1）配置上海本部 OSPF 进程号为 10，规划多区域。
（2）配置区域 0（S3、S4、EG1、EG2），VLAN 100 在区域 0 中发布。
（3）配置区域 1（S3、S4、S5）为完全 NSSA 类型。
（4）配置区域 2（S3、S4）。
（5）配置区域 3（S4、R1）
（6）配置 AC1、AC2 与 S3、S4 的静态路由。
（7）各设备禁止重分发直连，以 Network 发布明细网段。
（8）各分校 RIP 版本为 RIP-2，取消自动汇总功能。
（9）上海本部与分校通过重分发引入彼此路由。
（10）上海本部业务网段中不出现协议报文。
（11）所有路由协议都发布具体网段。
（12）为了管理方便，要求发布 Loopback 地址。
（13）优化 OSPF 相关配置，以尽量加快 OSPF 收敛。
（14）重分发路由进 OSPF 中使用类型 1。

6. 广域网链路配置与安全部署

上海本部路由器 R1 与北京校区路由器 R2、广州校区路由器 R3 之间属于广域网链路。其中，接入路由器 R1-R2 之间租用一条带宽为 2MB 的线路，接入路由器 R1 与 R3 间租用两条带宽均为 2MB 的线路。

上海本部的出口路由器与分校区的接入路由器之间使用广域网链路连接，使用 PPP 进行安全保护，同时提高接入路由器 R1 与 R3 的链路带宽，简化网络部署，具体要求如下。
（1）使用 CHAP 进行验证。
（2）使用双向认证方式，通过"用户名+验证口令"方式进行验证。
（3）配置验证的用户名和密码均为 ruijie。
（4）接入路由器 R1 与 R3 之间使用 PPP 链路捆绑，捆绑组号为 1。
（5）考虑到广域网线路安全性较差，所以使用 IPSec 对上海本部到各分校的数据流进行加密。
（6）要求使用动态隧道主模式，安全协议采用 AH-ESP，加密算法采用 3DES，认证算法采用 MD，以 IKE 方式建立 IPSec SA。
（7）在接入路由器 R1 上需要配置的参数要求如下。
① IPSec 加密转换集名称为 myset。
② 动态 IPSec 加密图名称为 dymymap。
③ 预共享密钥为明文 123456。
④ 静态的 IPSec 加密图名称为 mymap。
（8）在接入路由器 R2 和 R3 上需要配置的参数要求如下。
① ACL 编号为 101。
② 静态的 IPSec 加密图名称为 mymap。
③ 预共享密钥为明文 123456。

7. 路由选路部署

考虑到数据分流及负载均衡的目的,针对上海本部与各分校数据流的走向进行路由选路部署,要求如下。

(1)通过修改 OSPF 接口开销达到分流的目的,且其值必须为 5 或 10。
(2)OSPF 通过路由引入时,改变引入路由的开销值,且其值必须为 5 或 10。
(3)上海本部 VLAN 10、VLAN 20、VLAN 30 中的用户与互联网互通主路径规划为 S3-EG1。
(4)上海本部 VLAN 40 用户与互联网互通主路径规划为 S4-EG2。
(5)各分校用户与互联网互通主路径规划为 S4-EG2。
(6)主链路故障可无缝切换到备用链路上。
(7)来回数据流路径一致。

8. QoS 安全部署

为了防止大量不断突发的数据导致网络拥挤,必须对接入的用户流量加以限制。所配置的参数要求如下。

(1)在上海本部交换机 S1、S2 的 Gi0/1-16 接口处设置接口限速,限速 10Mbit/s。
(2)各分校接入路由器设备 R2、R3 做流量整形,Gi0/0 接口对接收到的报文进行流量控制,下行报文流量不能超过 1Mbit/s,如果超过流量限制,则将违规报文丢弃。

9. IPv6 部署

在核心交换机 S3、S4 上启用 IPv6 网络部署,实现 VLAN 10、VLAN 20、VLAN 30、VLAN 40 中的 IPv6 用户终端设备自动从网关处获取 IPv6 地址,并实现互连互通,实现网络资源共享。

在核心交换机 S3 和 S4 上配置 VRRP for IPv6,实现主机的 IPv6 网关冗余。
VRRP 与 MSTP 的主备状态与 IPv4 网络一致。
VRRP for IPv6 地址规划如表 37-6 所示。

表 37-6 VRRP for IPv6 地址规划

设备	接口	IPv6 地址	VRRP 组号	虚拟 IP 地址
S3	VLAN 10	2001:192:10::252/64	10	2001:192:10::254/64
	VLAN 20	2001:192:20::252/64	20	2001:192:20::254/64
	VLAN 30	2001:192:30::252/64	30	2001:192:30::254/64
	VLAN 40	2001:192:40::252/64	40	2001:192:40::254/64
S4	VLAN 10	2001:192:10::253/64	10	2001:192:10::254/64
	VLAN 20	2001:192:20::253/64	20	2001:192:20::254/64
	VLAN 30	2001:192:30::253/64	30	2001:192:30::254/64
	VLAN 40	2001:192:40::253/64	40	2001:192:40::254/64

【任务目标三】移动互联网搭建与网络优化

上海本部校区与各分校均需要规划和部署移动互联网，同时，为保证不同学生利用无线网络安全可靠地访问互联网，需要进行无线网络安全及性能优化配置，确保师生都有良好的上网体验。

1. 无线网络基础部署

配置 AC 作为上海本部网络中无线用户的 DHCP 服务器，使用交换机 S3、S4 为总部 AP 的 DHCP 服务器，其中，交换机 S3 分配 AP 地址范围为其网段的 1～100，交换机 S4 分配 AP 地址范围为其网段的 101～200；使用 S6、S7 交换机作为分校网络中无线用户与 AP 的 DHCP 服务器，为其终端自动分配地址。

创建上海本部 SSID(WLAN-ID 1)为 Ruijie-ZX_XX（XX 现场提供），AP-Group 为 ZX，上海本部无线用户关联 SSID 后可自动获取地址。

创建北京分校 SSID(WLAN-ID 2)为 Ruijie-BJFX_XX（XX 现场提供），AP-Group 为 BJFX，北京分校无线用户关联 SSID 后可自动获取地址。

创建广州分校 SSID(WLAN-ID 3)为 Ruijie-GZFX_XX（XX 现场提供），AP-Group 为 GZFX，广州分校无线用户关联 SSID 后可自动获取地址。

2. AC 集群部署

AC1 为主用，AC2 为备用。正常情况下，AP 与 AC1 建立隧道，当 AP 与 AC1 失去连接，等待超时后，会切换至 AC2 并提供服务。

3. 无线安全部署

上海本部无线用户接入无线网络时，需要采用本地 Web 认证方式，认证用户名密码为 XX（现场提供）。

为了防止无线局域网 ARP 欺骗影响用户上网体验，配置无线环境 ARP 欺骗防御功能。

在某些时候出于安全性的考虑，需要对在同一台 AP 中的用户彼此之间进行隔离，使用户之间不能互相访问，配置同一台 AP 下的用户间隔离功能。

4. 无线性能优化

要求在学校上海本部无线局域网的用户中启用集中转发模式，各分校区的无线网络用户启用本地转发模式。

限制每台 AP 关联的用户数最大为 16。

关闭低速率（1Mbit/s、6Mbit/s）应用接入。

【任务目标四】配置网络出口及网络优化

上海本部校区与分校无线用户需要通过独立的互联网线路访问外网资源，并针对访问资源进行用户身份认证与信息审计监督。

1. 出口 NAT 部署

在上海本部的网络出口网关上，限制无线局域网中的用户（ACL 编号为 110）只能通过联通线路访问互联网，通过 NAPT 转换方式接入互联网接口。

多层交换技术（实践篇）

上海本部网络对有线网络用户（ACL 编号为 111）不做限制，可访问联通网及教育网资源。

在上海本部网络的出口设备 EG1 上，配置上海本部网络中的核心交换机 S3（11.XX.0.3）设备开启 Telnet 服务，可以通过互联网进行访问，并将其地址映射至联通网线路上，映射地址为 196.XX.0.10（XX 现场提供）。

同时，需要确保 NAT 映射数据流来回一致，启用出口设备 EG 上的源进、源出功能，保证任何外网用户（联通、电信、移动、教育……）均可访问映射地址 196.XX.0.10。

2．全局流表策略部署

在用户没有实施防火墙限制的情况下，如果遇到大量的伪源 IP 地址攻击，或者端口扫描时，会把设备的流表占满，导致正常的数据无法建流而被丢弃。因此，要求部署全局流表防火墙，ACL（编号为 112）策略要求如下。

（1）允许所有 IP 地址到设备接口的 Web 管理和 Ping 操作。

（2）允许内网 IP 地址到外网所有资源的访问。

（3）允许任意 IP 地址访问映射的内网交换机上的资源。

3．应用流量控制部署

上海本部校区的网络需要配置针对访问外网的 SSH 流量限速，每个 IP 限速 1000kbit/s，内网 SSH 总流量不超过 100MB。

4．用户行为策略部署

上海本部校区的网络需要配置在工作时间（周一到周五上午 9:00～17:00），禁止员工使用 P2P 应用软件。

上海校区网络需要禁止内网用户通过浏览器访问 http://196.XX.0.2（XX 现场提供）。

5．数据分流与负载均衡

配置上海本部校区的网络与各分校用户数据流匹配，完成 EG 内置联通网与教育网地址库配置，实现访问联通网资源走联通网线路，访问教育网资源走教育网线路。

除联通网、教育网资源之外，默认所有数据流在联通网与教育网线路之间进行负载转发。

上海本部校区的联通网线路上每天 18:00～22:00 网络流量压力较大，配置在此时间段内将 P2P 应用软件流量引流到教育网线路。

【参考配置】

考虑到实际实施过程中现场设备数据的多少、模块的齐全程度、操作系统的版本差别、设备的连接接口信息、全网的地址规程状况以及读者掌握的相关知识多少各有不同，下面按照上述要求给出部分设备的参考配置。

1．上海本部校区中二层接入交换机 S1 设备配置参考

（1）使用"show run int Gi0/1"命令查看接口信息，如下所示。

```
S1#show run int gi0/1

Building configuration...
Current configuration: 234 bytes
```

任务 37 综合实训 2

```
 interface GigabitEthernet 0/1
  switchport protected
  description con_To_PC1
  switchport access vlan 10
  spanning-tree bpduguard enable
  spanning-tree portfast
  rate-limit output 10000 4096
  rldp port loop-detect shutdown-port
```

（2）使用"show run int Gi0/24"命令查看接口信息，如下所示。

```
S1#show run int gi0/24

Building configuration...
Current configuration: 251 bytes

interface GigabitEthernet 0/24
 description con_To_S4_Gi0/1
 switchport mode trunk
 no nfpp arp-guard enable
 no nfpp icmp-guard enable
 no nfpp ip-guard enable
 no nfpp dhcp-guard enable
 no nfpp dhcpv6-guard enable
 no nfpp nd-guard enable
```

（3）使用"show run | in err"命令查看配置信息中的恢复信息，如下所示。

```
S1(config)#show run | in err
errdisable recovery interval 300
```

（4）使用"show access-lists"命令，查看访问控制列表信息，如下所示。

```
S1#show access-lists

ip access-list extended 101
 10 deny udp any eq bootps any
 20 permit ip any any
```

（5）使用"show spanning-tree summary"命令，查看生成树信息，如下所示。

```
S1#show spanning-tree summary

Spanning tree enabled protocol mstp
MST 0 vlans map : 1-9, 11-19, 21-29, 31-39, 41-49, 51-99, 101-4094
   Root ID    Priority    4096
              Address     5869.6cf5.224c
              this bridge is root
              Hello Time  2 sec  Forward Delay 15 sec  Max Age 20 sec

   Bridge ID  Priority    32768
              Address     5869.6cd5.b88b
              Hello Time  2 sec  Forward Delay 15 sec  Max Age 20 sec

Interface        Role Sts Cost       Prio     OperEdge Type
---------------- ---- --- ---------- -------- -------- ----------------
Gi0/24           Altn BLK 20000      128      False    P2p
Gi0/23           Root FWD 20000      128      False    P2p

MST 1 vlans map : 10, 20, 30
   Region Root Priority    4096
              Address     5869.6cf5.224c
              this bridge is region root

   Bridge ID  Priority    32768
              Address     5869.6cd5.b88b

Interface        Role Sts Cost       Prio     OperEdge Type
---------------- ---- --- ---------- -------- -------- ----------------
Gi0/24           Altn BLK 20000      128      False    P2p
Gi0/23           Root FWD 20000      128      False    P2p

MST 2 vlans map : 40, 50, 100
   Region Root Priority    4096
              Address     5869.6cf5.236c
              this bridge is region root

   Bridge ID  Priority    32768
              Address     5869.6cd5.b88b

Interface        Role Sts Cost       Prio     OperEdge Type
---------------- ---- --- ---------- -------- -------- ----------------
Gi0/24           Root FWD 20000      128      False    P2p
Gi0/23           Altn BLK 20000      128      False    P2p
```

多层交换技术（实践篇）

2. 上海本部校区中的二层接入交换机 S2 设备配置参考

使用"show version"命令，查看交换机版本信息，如下所示。

```
S2#show version
System description         : Ruijie 10G Ethernet Switch(S2910-24GT4XS-E)
System start time          : 2018-01-12 09:31:33
System uptime              : 0:10:00:33
System hardware version    : 1.10
System software version    : S2910_RGOS 11.4(1)B1P3
System patch number        : NA
System serial number       : G1LD2ES004118
System boot version        : 1.2.13
Module information:
   Slot 0 : S2910-24GT4XS-E
     Hardware version      : 1.10
     Boot version          : 1.2.13
     Software version      : S2910_RGOS 11.4(1)B1P3
     Serial number         : G1LD2ES004118
```

3. 上海本部校区核心交换机 S3 设备配置参考

（1）使用"show run | be dhcp"命令，查看 DHCP 信息，如下所示。

```
S3(config)#show run | be dhcp
service dhcp
!
ip dhcp pool vlan10
 network 192.1.10.0 255.255.255.0
 default-router 192.1.10.254
!
ip dhcp pool vlan50
 option 138 ip 11.1.0.204 11.1.0.205
 network 192.1.50.0 255.255.255.0 192.1.50.1 192.1.50.100
 default-router 192.1.50.254
```

（2）使用"show spanning-tree summary"命令，查看生成树信息，如下所示。

```
S3#show spanning-tree summary
Spanning tree enabled protocol mstp
MST 0 vlans map : 1-9, 11-19, 21-29, 31-39, 41-49, 51-99, 101-4094
    Root ID     Priority    4096
                Address     5869.6cf5.224c
                this bridge is root
                Hello Time   2 sec  Forward Delay 15 sec  Max Age 20 sec

    Bridge ID   Priority    4096
                Address     5869.6cf5.224c
                Hello Time   2 sec  Forward Delay 15 sec  Max Age 20 sec

Interface           Role Sts Cost        Prio      OperEdge  Type
----------------    ---- --- ----------  --------  --------  ----------------
Ag1                 Desg FWD 1900        128       False     P2p
Gi0/3               Desg FWD 20000       128       False     P2p
Gi0/2               Desg FWD 20000       128       False     P2p
Gi0/1               Desg FWD 20000       128       False     P2p

MST 1 vlans map : 10, 20, 30
    Region Root Priority    4096
                Address     5869.6cf5.224c
                this bridge is region root

    Bridge ID   Priority    4096
                Address     5869.6cf5.224c

Interface           Role Sts Cost        Prio      OperEdge  Type
----------------    ---- --- ----------  --------  --------  ----------------
Ag1                 Desg FWD 1900        128       False     P2p
Gi0/3               Desg FWD 20000       128       False     P2p
Gi0/2               Desg FWD 20000       128       False     P2p
Gi0/1               Desg FWD 20000       128       False     P2p

MST 2 vlans map : 40, 50, 100
    Region Root Priority    4096
                Address     5869.6cf5.236c
                this bridge is region root

    Bridge ID   Priority    8192
                Address     5869.6cf5.224c

Interface           Role Sts Cost        Prio      OperEdge  Type
----------------    ---- --- ----------  --------  --------  ----------------
Ag1                 Root FWD 1900        128       False     P2p
Gi0/3               Desg FWD 20000       128       False     P2p
Gi0/2               Desg FWD 20000       128       False     P2p
Gi0/1               Desg FWD 20000       128       False     P2p
```

任务 37 综合实训 2

（3）使用"show vrrp brief"命令，查看 VRRP 信息，如下所示。

```
S3#show vrrp brief
Interface        Grp   Pri   timer   Own   Pre   State    Master addr
VLAN 10          10    150   3.41    -     P     Master   192.1.10.252
VLAN 20          20    150   3.41    -     P     Master   192.1.20.252
VLAN 30          30    150   3.41    -     P     Master   192.1.30.252
VLAN 40          40    120   3.53    -     P     Backup   192.1.40.253
VLAN 50          50    120   3.53    -     P     Backup   192.1.50.253
VLAN 100         100   120   3.53    -     P     Backup   192.1.100.253
```

（4）使用"show run | be ospf 10"命令，查看 OSPF 信息，如下所示。

```
S3#show run | be ospf 10
router ospf 10
 graceful-restart
 redistribute static metric-type 1 subnets
 passive-interface VLAN 10
 passive-interface VLAN 20
 passive-interface VLAN 30
 passive-interface VLAN 40
 passive-interface VLAN 50
 area 1 nssa no-summary
 network 10.1.0.0 0.0.0.3 area 1
 network 10.1.0.4 0.0.0.3 area 0
 network 10.1.0.40 0.0.0.3 area 0
 network 11.1.0.33 0.0.0.0 area 2
 network 192.1.10.0 0.0.0.255 area 2
 network 192.1.20.0 0.0.0.255 area 2
 network 192.1.30.0 0.0.0.255 area 2
 network 192.1.40.0 0.0.0.255 area 2
 network 192.1.50.0 0.0.0.255 area 2
 network 192.1.100.0 0.0.0.255 area 0
```

（5）使用"show ip ospf neighbor"命令，查看 OSPF 邻居信息，如下所示。

```
S3#show ip ospf neighbor
OSPF process 10, 4 Neighbors, 4 is Full:
Neighbor ID    Pri   State      BFD State   Dead Time   Address
11.1.0.5       1     Full/ -    -           00:00:33    10.1.0.2
11.1.0.11      1     Full/ -    -           00:00:34    10.1.0.6
11.1.0.12      1     Full/ -    -           00:00:30    10.1.0.42
11.1.0.34      1     Full/DR    -           00:00:39    192.1.100.253
```

（6）使用"show ip route ospf"命令，查看 OSPF 路由信息，如下所示。

```
S3#show ip route ospf
O*E2  0.0.0.0/0 [110/1] via 10.1.0.6, 04:21:41, GigabitEthernet 0/6
O     10.1.0.8/30 [110/6] via 10.1.0.42, 00:00:55, GigabitEthernet 0/4
                  [110/6] via 192.1.100.253, 00:00:55, VLAN 100
O IA  10.1.0.12/30 [110/6] via 192.1.100.253, 00:00:55, VLAN 100
O     10.1.0.32/30 [110/2] via 10.1.0.2, 01:49:17, GigabitEthernet 0/5
O     10.1.0.36/30 [110/6] via 10.1.0.6, 00:00:55, GigabitEthernet 0/6
O IA  11.1.0.1/32 [110/6] via 192.1.100.253, 00:00:55, VLAN 100
O     11.1.0.5/32 [110/1] via 10.1.0.2, 01:49:17, GigabitEthernet 0/5
O     11.1.0.11/32 [110/1] via 10.1.0.6, 00:00:48, GigabitEthernet 0/6
O IA  11.1.0.12/32 [110/5] via 10.1.0.42, 00:00:55, GigabitEthernet 0/4
O IA  11.1.0.34/32 [110/5] via 192.1.100.253, 00:00:55, VLAN 100
O N1  172.16.0.0/24 [110/21] via 10.1.0.2, 00:15:27, GigabitEthernet 0/5
O E1  194.1.10.0/24 [110/26] via 192.1.100.253, 00:00:55, VLAN 100
O E1  194.1.20.0/24 [110/26] via 192.1.100.253, 00:00:55, VLAN 100
O E1  194.1.30.0/24 [110/26] via 192.1.100.253, 00:00:55, VLAN 100
O E1  195.1.10.0/24 [110/26] via 192.1.100.253, 00:00:55, VLAN 100
O E1  195.1.20.0/24 [110/26] via 192.1.100.253, 00:00:55, VLAN 100
O E1  195.1.30.0/24 [110/26] via 192.1.100.253, 00:00:55, VLAN 100
```

（7）使用"show ipv6 interface brief"命令，查看 IPv6 摘要信息，如下所示。

```
S3#show ipv6 interface brief
VLAN 10                              [up/up]
    2001:198:10::252
    FE80::5A69:6CFF:FEF5:224D
VLAN 20                              [up/up]
    2001:198:20::252
    FE80::5A69:6CFF:FEF5:224D
VLAN 30                              [up/up]
```

```
                    2001:198:30::252
                    FE80::5A69:6CFF:FEF5:224D
VLAN 40                                              [up/up]
                    2001:198:40::252
                    FE80::5A69:6CFF:FEF5:224D
```

（8）使用"show interfaces switchport | in Agg"命令，查看聚合接口信息，如下所示。

```
S3(config)#show interfaces switchport | in Agg
AggregatePort 1              enabled    TRUNK    1    1
                                                      Disabled  10,20,30,40,50,60,100
```

4. 上海本部校区核心交换机 S4 设备配置参考

（1）使用"show ip dhcp binding"命令，查看 DHCP 地址获取信息，如下所示。需要将计算机连接到接入交换机，以获取 IP 地址。

```
S4#show ip dhcp binding

Total number of clients   : 1
Expired clients           : 0
Running clients           : 1

IP address        Hardware address      Lease expiration              Type
192.1.50.101      5869.6c51.015a        000 days 21 hours 31 mins    Automatic
```

（2）使用"show ip ospf neighbor"命令，查看 OSPF 邻居信息，如下所示。

```
S4#show ip ospf neighbor

OSPF process 10, 5 Neighbors, 5 is Full:
Neighbor ID   Pri   State      BFD State   Dead Time   Address        Interface
11.1.0.12     1     Full/-      -          00:00:34    10.1.0.10      GigabitEthernet 0/6
11.1.0.1      1     Full/-      -          00:00:31    10.1.0.14      GigabitEthernet 0/7
11.1.0.5      1     Full/-      -          00:00:31    10.1.0.34      GigabitEthernet 0/5
11.1.0.11     1     Full/-      -          00:00:35    10.1.0.38      GigabitEthernet 0/4
11.1.0.33     1     Full/BDR    -          00:00:32    192.1.100.252  VLAN 100
```

（3）使用"show ip route static"命令，查看静态路由信息，如下所示。

```
S4#show ip route static
S    11.1.0.204/32 [1/0] via 192.1.100.2
S    11.1.0.205/32 [1/0] via 192.1.100.3
S    192.1.60.0/24 [1/0] via 192.1.100.1
```

（4）使用"show ipv6 vrrp brief"命令，查看 IPv6 的 VRRP 信息，如下所示。

```
S4#show ipv6 vrrp brief
Interface    Grp   Pri   timer   Own   Pre   State    Master addr
VLAN 10      10    120   3.53    -     P     Backup   FE80::5A69:6CFF:FEF5:224D
VLAN 20      20    120   3.53    -     P     Backup   FE80::5A69:6CFF:FEF5:224D
VLAN 30      30    120   3.53    -     P     Backup   FE80::5A69:6CFF:FEF5:224D
VLAN 40      40    255   3.00    O     P     Master   FE80::5A69:6CFF:FEF5:236D
                                                      Group addr
                                                      FE80::5A69:6CFF:FEF5:224D
                                                      FE80::5A69:6CFF:FEF5:224D
                                                      FE80::5A69:6CFF:FEF5:224D
                                                      FE80::5A69:6CFF:FEF5:236D
```

5. 上海本部校区核心网络中的三层交换机 S5 设备配置参考

（1）使用"show run | in password""show run | be line"命令，查看配置文件密码信息，如下所示。

```
S5#show run | in password
username admin password admin
no service password-encryption
enable password admin
S5#show run | be line
line console 0
line vty 0 4
 login local
!
end
```

（2）使用"show run | in snmp"命令，查看 SNMP 管理信息，如下所示。

```
S5#show run | in snmp
snmp-server host 172.16.0.254 traps version 2c ruijie
snmp-server host 172.16.0.254 traps version 2c public
snmp-server enable traps
snmp-server community ruijie rw
snmp-server community public ro
```

（3）使用 "show ip deny" 命令，查看安全信息，如下所示。

```
S5#show ip deny
  DoS Protection Mode                          State
  -----------------------------------          -----
  Protect against Land attack                  On
  Protect against invalid L4port attack        On
  Protect against invalid TCP attack           On
```

6. 北京分校区核心网络中的三层交换机 S6 设备配置参考

（1）使用 "show run int Gi0/1" 命令，查看接口信息，如下所示。

```
S6#show run int gi0/1
Building configuration...
Current configuration: 180 bytes

interface GigabitEthernet 0/1
 description con_To_PC2
 switchport access vlan 10
 spanning-tree bpduguard enable
 spanning-tree portfast
 rldp port loop-detect shutdown-port
```

（2）使用 "show ip dhcp binding" 命令，查看 DHCP 地址信息，如下所示。

```
S6#show ip dhcp binding
Total number of clients   : 3
Expired clients           : 0
Running clients           : 3

IP address      Hardware address      Lease expiration
194.1.20.2      f816.54c0.9289        000 days 22 hours 52 mins
194.1.20.1      0446.657c.b926        000 days 22 hours 21 mins
194.1.30.1      5869.6c51.8b46        000 days 19 hours 19 mins
```

7. 广州分校区核心网络中三层交换机 S7 设备配置参考

（1）使用 "show ip int brief" 命令，查看 IP 摘要信息，如下所示。

```
S7#show ip int brief
Interface              IP-Address(Pri)     IP-Address(Sec)    Status
GigabitEthernet 0/24   10.1.0.30/30        no address         up
Loopback 0             11.1.0.7/32         no address         up
VLAN 10                195.1.10.254/24     no address         up
VLAN 20                195.1.20.254/24     no address         up
VLAN 30                195.1.30.254/24     no address         up
```

（2）使用 "show vlan" 命令，查看 VLAN 信息，如下所示。

```
S7#show vlan
VLAN Name              Status     Ports
---- ---------------   --------   --------------------------------
   1 VLAN0001          STATIC     Gi0/21, Gi0/22, Gi0/23, Te0/25
                                  Te0/26, Te0/27, Te0/28
  10 Wire_user         STATIC     Gi0/1, Gi0/2, Gi0/3, Gi0/4
                                  Gi0/5, Gi0/6, Gi0/7, Gi0/8
                                  Gi0/9, Gi0/10, Gi0/11, Gi0/12
                                  Gi0/13, Gi0/14, Gi0/15, Gi0/16
                                  Gi0/17, Gi0/18, Gi0/19, Gi0/20
                                  Gi0/21, Gi0/22, Gi0/23
  20 Wireless_user     STATIC     Gi0/21, Gi0/22, Gi0/23
  30 AP                STATIC     Gi0/21, Gi0/22, Gi0/23
```

8. 上海本部校区出口路由器 R1 设备配置参考

（1）使用 "show run | include ssh" "show run | include username" "show run | be line vty" 命令，查看路由器配置信息，如下所示。

```
R1#show run | include ssh
enable service ssh-server
R1#show run | inc
R1#show run | include username
webmaster level 0 username admin password 7 0121474e3e16
username admin password admin
username ruijie password ruijie
R1#show run | be line vty
line vty 0 4
 login local
```

（2）使用"show run int s 2/0"命令，查看配置信息，如下所示。

```
R1#show run int s2/0

Building configuration...
Current configuration : 228 bytes
!
interface Serial 2/0
 encapsulation PPP
 ppp authentication chap
 ppp chap hostname ruijie
 ppp chap password ruijie
 ip address 10.1.0.18 255.255.255.252
 crypto map mymap
 clock rate 64000
 description con_To_R2_S2/0
```

（3）使用"show ip int brief"命令，查看 IP 摘要信息，如下所示。

```
R1#show ip int brief
Interface              IP-Address(Pri)    IP-Address(Sec)    Status
multilink 1            10.1.0.22/30       no address         up
Serial 2/0             10.1.0.18/30       no address         up
Serial 3/0             no address         no address         up
Serial 4/0             no address         no address         up
GigabitEthernet 0/0    10.1.0.14/30       no address         up
GigabitEthernet 0/1    no address         no address         up
Loopback 0             11.1.0.1/32        no address         up
```

（4）使用"show ppp multilink"命令，查看 PPP 信息，如下所示。

```
R1#show ppp multilink

multilink 1 (active)
  authname ( ruijie ) / endpoint( ruijie )
  interface state: UP
  ipcp state: Open
  ipv6cp state: not Open
  frag queue: 0 bytes, 0 frags,  drops: 0 timeout, 0 lack
  group members: 2, active members: 2, bund min/max: 0/16
      Serial 4/0 (active)
      Serial 3/0 (active)

1 MLP bundle in system
```

（5）使用"show crypto isakmap sa"命令，查看加密信息，如下所示。

```
R1#show crypto isakmp sa
destination     source          state
10.1.0.21       10.1.0.22       IKE_IDLE
10.1.0.17       10.1.0.18       IKE_IDLE
         conn-id        lifetime(second)
         0              69665
         1              69667
```

9. 北京分校区接入路由器 R2 设备配置参考

（1）使用"show crypto isakmap sa"命令，查看加密信息，如下所示。

```
R2#show crypto isakmap sa
destination     source          state
10.1.0.18       10.1.0.17       IKE_IDLE
         conn-id        lifetime(second)
         0              83598
```

（2）使用"show crypto ipsec sa | include pkts"命令，查看安全加密信息，如下所示。

```
R2(config)#show crypto ipsec sa | include pkts
    #pkts encaps: 0, #pkts encrypt: 0, #pkts digest 0
    #pkts decaps: 0, #pkts decrypt: 0, #pkts verify 0
    #pkts encaps: 94, #pkts encrypt: 94, #pkts digest 94
    #pkts decaps: 100, #pkts decrypt: 100, #pkts verify 100
    #pkts encaps: 64, #pkts encrypt: 64, #pkts digest 64
    #pkts decaps: 72, #pkts decrypt: 72, #pkts verify 72
```

（3）使用"show crypto ipsec transform-set"命令，查看安全信息，如下所示。

```
R2#show crypto ipsec transform-set
transform set myset: { ah-md5-hmac,esp-md5-hmac,esp-3des,}
        will negotiate = {Tunnel,}
```

（4）使用"show ip route"命令，查看路由表信息，如下所示。

```
R2#show ip route
Codes: C - connected, S - static, R - RIP, B - BGP
       O - OSPF, IA - OSPF inter area
       N1 - OSPF NSSA external type 1, N2 - OSPF NSSA external type 2
       E1 - OSPF external type 1, E2 - OSPF external type 2
       i - IS-IS, su - IS-IS summary, L1 - IS-IS level-1, L2 - IS-IS level-2
       ia - IS-IS inter area, * - candidate default

Gateway of last resort is 10.1.0.18 to network 0.0.0.0
S*    0.0.0.0/0 [1/0] via 10.1.0.18
C     10.1.0.16/30 is directly connected, Serial 2/0
C     10.1.0.17/32 is local host.
C     10.1.0.18/32 is directly connected, Serial 2/0
C     10.1.0.24/30 is directly connected, GigabitEthernet 0/0
C     10.1.0.25/32 is local host.
C     11.1.0.2/32 is local host.
R     11.1.0.6/32 [120/1] via 10.1.0.26, 04:47:49, GigabitEthernet 0/0
R     194.1.10.0/24 [120/1] via 10.1.0.26, 04:46:00, GigabitEthernet 0/0
R     194.1.20.0/24 [120/1] via 10.1.0.26, 04:46:00, GigabitEthernet 0/0
R     194.1.30.0/24 [120/1] via 10.1.0.26, 04:45:59, GigabitEthernet 0/0
```

10. 广州分校区接入路由器 R3 设备配置参考

（1）使用"show rate-limit interface Gi0/0"命令，查看接口限速信息，如下所示。

```
R3#show rate-limit interface gi0/0
GigabitEthernet 0/0
  Input
    matches all traffic
      params:  1000000 bps, 100000 limit, 200000 extended limit
      conformed 4496 packets, 990357 bytes; action: transmit
      exceeded 0 packets, 0 bytes; action: drop
      cbucket 299873. cbs 300000; ebucket 0 ebs 0
```

（2）使用"show run int multilink 1"命令，查看配置信息，如下所示。

```
R3#show run int multilink 1

Building configuration...
Current configuration : 155 bytes
!
interface multilink 1
 ip address 10.1.0.21 255.255.255.252
 crypto map mymap
 traffic-shape rate 4000000 80000 80000 1000
 max-reserved-bandwidth 99
```

（3）使用"show access-lists"命令，查看访问控制列表信息，如下所示。

```
R3#show access-lists

ip access-list extended 101
 10 permit ip 195.1.0.0 0.0.255.255 192.1.0.0 0.0.255.255
 20 permit ip 195.1.0.0 0.0.255.255 host 11.1.0.204
 30 permit ip 195.1.0.0 0.0.255.255 host 11.1.0.205
```

11. 上海本部校区无线控制器 AC1 设备配置参考

（1）使用"show version all"命令，查看版本信息，如下所示。

```
AC1#show version all
AP(AP-1)'s version:
  Product ID         : AP3220
  System uptime      : 0:4:42:58
  Hardware version   : 1.01
  Software version   : AP_RGOS 11.1(5)B81P3, Release(03241713)
  Patch number       : NA
  Software number    : M13253212172016
  Serial number      : G1JDA1K475050
  MAC address        : 5869.6c51.0159

AP(AP-2)'s version:
  Product ID         : AP3220
  System uptime      : 0:4:42:58
  Hardware version   : 1.01
  Software version   : AP_RGOS 11.1(5)B81P3, Release(03241713)
  Patch number       : NA
  Software number    : M13253212172016
  Serial number      : G1JDB1K000638
  MAC address        : 5869.6c51.8919
```

（2）使用"show run int Gi0/1"命令，查看接口信息，如下所示。

```
AC1#show run int gi0/1

Building configuration...
Current configuration: 129 bytes

interface GigabitEthernet 0/1
 description con_To_S3_Gi0/3
 switchport mode trunk
 switchport trunk allowed vlan only 60,100
```

（3）使用"show ip dhcp binding"命令，查看地址获取信息，如下所示。

```
AC1#show ip dhcp binding

Total number of clients    : 2
Expired clients            : 1
Running clients            : 1

IP address        Hardware address      Lease expiration
192.1.60.2        f816.54c0.9289        000 days 23 hours 46 mins
```

（4）使用"show ap-group aps summary"命令，查看AP组信息，如下所示。

```
AC1#show ap-group aps summary
Ap Name                                    AP group
-------                                    --------
AP-2                                       bjfx
AP-3                                       gzfx
AP-1                                       zx
                                                                        ctx_name
                                                                        ----------
                                                                        11.1.0.205-10
                                                                        11.1.0.205-10
                                                                        11.1.0.205-10
```

（5）使用"show ap-config summary"命令，查看AC状态信息，如下所示。

```
AC1#show ap-config summary
========= show ap status =========
Radio: Radio ID or Band: 2.4G = 1#, 5G = 2#
       E = enabled, D = disabled, N = Not exist
       Current Sta number
       Channel: * = Global
       Power Level = Percent

Online AP number: 3
Offline AP number: 0

AP Name                         IP Address      Mac Address
-------                         ----------      -----------
AP-1                            192.1.50.1      5869.6c51.0159
AP-2                            194.1.30.1      5869.6c51.8b45
AP-3                            195.1.30.1      5869.6c51.8bbd
Radio           Radio                    Up/Off time     State
-----           -----                    ----------      -----
1  E  0  6*  100  2  E  0  165*  100    0:00:40:42     Run
1  E  0  1*  100  2  E  0  161*  100    0:00:50:35     Run
1  E  0  11* 100  2  E  0  157*  100    0:00:54:34     Run
```

（6）使用"show ap-config running"命令，查看AC配置信息，如下所示。

```
AC1#show ap-config running

Building configuration...
Current configuration: 328 bytes

!
ap-config AP-1
 ap-mac 5869.6c51.0159
 ap-group zx
 sta-limit 16
 no 11acsupport enable radio 2
!
ap-config AP-2
 ap-mac 5869.6c51.8b45
 ap-group bjfx
 sta-limit 16
 no 11acsupport enable radio 2
!
ap-config AP-3
 ap-mac 5869.6c51.8bbd
 ap-group gzfx
 sta-limit 16
 no 11acsupport enable radio 2
!!!!!
```

任务 ㊲ 综合实训 2

（7）使用"show web-auth user all"命令，查看 web 认证信息，如下所示。

```
AC1#show web-auth user all
Current user num: 1, Online 1
Address                                              Online   Time Limit         Time used
---------------------------------------------------  ------   ---------------    ---------------
192.1.60.2                                           On       0d 00:00:00        0d 00:54:21
  Status              Name
  ---------------     ---------
  Active              admin
```

（8）使用"show run | be wids"命令，查看配置信息，如下所示。

```
AC1(config)#show run | be wids
wids
 user-isolation ap enable
```

（9）使用"show wlan-config cb 2"命令，查看无线配置信息，如下所示。

```
AC1#show wlan-config cb 2
WLAN ID........................................ 2
SSID........................................... Ruijie-BJFX-01
Profile........................................
MAC Mode....................................... Local
Tunnel Mode.................................... Local Bridging
Suppress SSID.................................. Disable
Sta-limit...................................... 0
NAS ID.........................................
Band Select.................................... Disable
SSID Code......................................
```

（10）使用"show wlan hot-backup 11.1.0.205"命令，查看无线热备信息，如下所示。

```
AC1#show wlan hot-backup 11.1.0.205
wlan hot-backup 11.1.0.205
  hot-backup      : Enable
  connect state   : CHANNEL_UP
  hello-interval  : 1000
  kplv-pkt        : ip
  work-mode       : NORMAL
  !
  context 10
    hot-backup role         : PAIR-ACTIVE
    hot-backup rdnd state   : REALTIME-SYN
    hot-backup priority     : 7
```

（11）使用"show wlan arp-check list"命令，查看无线安全检查信息，如下所示。

```
AC1#show wlan arp-check list
INTERFACE           SENDER MAC          SENDER IP           POLICY SOURCE
---------------     ---------------     ---------------     ---------------
Wlan 3              f816.54c0.9289      195.1.20.1          DHCP snooping
```

（12）使用"show run | be ac-c"命令，查看 AC 的配置信息，如下所示。

```
AC1#show run | be ac-c
ac-controller
  ac-name Ruijie_Ac
  country CN
  802.11g network rate 1 disabled
  802.11g network rate 2 disabled
  802.11g network rate 5 disabled
  802.11g network rate 6 disabled
  802.11g network rate 9 supported
  802.11g network rate 11 mandatory
  802.11g network rate 12 supported
  802.11g network rate 18 supported
  802.11g network rate 24 supported
  802.11g network rate 36 supported
  802.11g network rate 48 supported
  802.11g network rate 54 supported
```

12. 上海本部校区无线控制器 AC2 设备配置参考

（1）使用"show wlan hot-backup 11.1.0.204"命令，查看无线热备信息，如下所示。

```
AC2#show wlan hot-backup 11.1.0.204
wlan hot-backup 11.1.0.204
  hot-backup      : Enable
  connect state   : CHANNEL_UP
  hello-interval  : 1000
  kplv-pkt        : ip
  work-mode       : NORMAL
  !
```

```
context 10
  hot-backup role         : PAIR-STANDBY
  hot-backup rdnd state   : REALTIME-SYN
  hot-backup priority     : 4
```

（2）使用"show vrrp brief"命令，查看 VRRP 路由信息，如下所示。

```
AC2#show vrrp brief
Interface       Grp  Pri  timer  Own  Pre  State   Master addr
VLAN 60         1    100  3.60   -    P    Backup  192.1.60.252
VLAN 100        1    100  3.60   -    P    Backup  192.1.100.2

                                                    Group addr
                                                    192.1.60.254
                                                    192.1.100.1
```

（3）使用"show ac-config client"命令，查看 AC 连接客户端信息，如下所示。

```
AC2#show ac-config client
========= show sta status =========
AP     : ap name/radio id
Status : Speed/Power Save/Work Mode/Roaming State/MU MIMO, E = enable power save, D = disable power

Total Sta Num : 2
STA MAC         IPV4 Address    AP                              Wlan Vlan  Status
--------------  --------------  ------                          ---- ----  ------
0446.657c.b926  195.1.20.1      AP-2/1                          2    20    65.0M/E/bgn
7802.f8f0.012b  192.1.60.2      AP-1/2                          1    60    72.5M/D/an

Asso Auth       Net Auth        Up time
--------------  --------------  ------
OPEN            OPEN            0:02:29:39
OPEN            OPEN            0:00:00:25
```

（4）使用"show ap-config summary"命令，查看 AC 配置信息，如下所示。

```
AC2#show ap-config summary
========= show ap status =========
Radio: Radio ID or Band: 2.4G = 1#, 5G = 2#
       E = enabled, D = disabled, N = Not exist
       Current Sta number
       Channel: * = Global
       Power Level = Percent

Online AP number: 3
Offline AP number: 0

AP Name                                 IP Address      Mac Address
--------                                ----------      -----------
AP-1                                    192.1.50.1      5869.6c51.0159
AP-2                                    194.1.30.1      5869.6c51.8b45
AP-3                                    195.1.30.1      5869.6c51.8bbd
Radio           Radio                   Up/Off time     State
-----           -----                   -----------     -----
1 E  0  6*  100 2 E  0  165*  100       0:01:20:31      Run
1 E  0  1*  100 2 E  0  161*  100       0:01:27:12      Run
1 E  0  11* 100 2 E  0  157*  100       0:01:27:13      Run
```

（5）使用"show interfaces description"命令，查看接口描述信息，如下所示。

```
AC2#show interfaces description
Interface           Status    Administrative   Description
---------           ------    --------------   -----------
GigabitEthernet 0/1  up        up               con_To_S4_Gi0/3
```

13. 上海本部校区出口设备 EG1 配置参考

（1）使用"show ip route ospf"命令，查看 OSPF 路由信息，如下所示。

```
EG1#show ip route ospf
O IA   10.1.0.0/30   [110/2] via 10.1.0.5,  02:07:41, GigabitEthernet 0/0
O      10.1.0.8/30   [110/6] via 10.1.0.37, 00:19:05, GigabitEthernet 0/3
O IA   10.1.0.12/30  [110/6] via 10.1.0.37, 00:19:05, GigabitEthernet 0/3
O      10.1.0.32/30  [110/3] via 10.1.0.5,  02:07:26, GigabitEthernet 0/0
O      10.1.0.40/30  [110/6] via 10.1.0.5,  03:07:11, GigabitEthernet 0/0
O IA   11.1.0.1/32   [110/2] via 10.1.0.37, 00:19:05, GigabitEthernet 0/3
O IA   11.1.0.5/32   [110/2] via 10.1.0.5,  02:07:26, GigabitEthernet 0/0
O      11.1.0.12/32  [110/6] via 10.1.0.5,  00:19:05, GigabitEthernet 0/0
                     [110/6] via 10.1.0.37, 00:19:05, GigabitEthernet 0/3
```

```
O IA  11.1.0.33/32  [110/1]  via 10.1.0.5,  00:23:33, GigabitEthernet 0/0
O IA  11.1.0.34/32  [110/5]  via 10.1.0.37, 00:19:05, GigabitEthernet 0/3
O E1  11.1.0.204/32 [110/26] via 10.1.0.5,  00:18:58, GigabitEthernet 0/0
O E1  11.1.0.205/32 [110/26] via 10.1.0.5,  00:18:58, GigabitEthernet 0/0
O IA  192.1.10.0/24 [110/6]  via 10.1.0.5,  00:18:58, GigabitEthernet 0/0
O IA  192.1.20.0/24 [110/6]  via 10.1.0.5,  00:18:58, GigabitEthernet 0/0
O IA  192.1.30.0/24 [110/6]  via 10.1.0.5,  00:18:58, GigabitEthernet 0/0
O IA  192.1.40.0/24 [110/10] via 10.1.0.37, 00:19:05, GigabitEthernet 0/3
O IA  192.1.50.0/24 [110/2]  via 10.1.0.5,  03:07:11, GigabitEthernet 0/0
O E1  192.1.60.0/24 [110/26] via 10.1.0.5,  00:18:58, GigabitEthernet 0/0
O     192.1.100.0/24 [110/6] via 10.1.0.5,  00:18:58, GigabitEthernet 0/0
O E1  194.1.10.0/24 [110/26] via 10.1.0.37, 00:19:05, GigabitEthernet 0/3
O E1  194.1.20.0/24 [110/26] via 10.1.0.37, 00:19:05, GigabitEthernet 0/3
O E1  194.1.30.0/24 [110/26] via 10.1.0.37, 00:19:05, GigabitEthernet 0/3
O E1  195.1.10.0/24 [110/26] via 10.1.0.37, 00:19:05, GigabitEthernet 0/3
O E1  195.1.20.0/24 [110/26] via 10.1.0.37, 00:19:07, GigabitEthernet 0/3
O E1  195.1.30.0/24 [110/26] via 10.1.0.37, 00:19:07, GigabitEthernet 0/3
```

（2）使用"show run | be ip nat pool"命令，查看 NAT 地址转换信息，如下所示。

```
EG1#show run | be ip nat pool
ip nat pool wuxian prefix-length 24
 address interface GigabitEthernet 0/1 match interface GigabitEthernet 0/1
!
ip nat pool youxian prefix-length 24
 address interface GigabitEthernet 0/1 match interface GigabitEthernet 0/1
 address interface GigabitEthernet 0/2 match interface GigabitEthernet 0/2
!
ip nat pool nat_pool prefix-length 24
 address interface GigabitEthernet 0/1 match interface GigabitEthernet 0/1
 address interface GigabitEthernet 0/2 match interface GigabitEthernet 0/2
!
ip nat inside source static tcp 11.1.0.3 23 196.1.0.10 23 permit-inside
ip nat inside source list 110 pool wuxian overload
ip nat inside source list 111 pool youxian overload
```

（3）使用"show access-lists"命令，查看访问控制列表信息，如下所示。

```
EG1#show access-lists

ip access-list extended 110
 10 permit ip 192.1.60.0 0.0.0.255 any
 20 permit ip 194.1.20.0 0.0.0.255 any
 30 permit ip 195.1.20.0 0.0.0.255 any

ip access-list extended 111
 10 permit ip 192.1.10.0 0.0.0.255 any
 20 permit ip 192.1.20.0 0.0.0.255 any
 30 permit ip 192.1.30.0 0.0.0.255 any
 40 permit ip 192.1.40.0 0.0.0.255 any
 50 permit ip 194.1.10.0 0.0.0.255 any
 60 permit ip 195.1.10.0 0.0.0.255 any

ip access-list extended 112
 20 permit tcp any any eq www
 30 permit icmp any any
 40 permit ip 192.1.0.0 0.0.255.255 any
 50 permit ip 194.1.0.0 0.0.255.255 any
 60 permit ip 195.1.0.0 0.0.255.255 any
 70 permit tcp any host 196.1.0.10 eq telnet
 80 permit ospf any any
```

（4）使用"show run int Gi0/3"命令，查看接口信息，如下所示。

```
EG1#show run int gi0/3

Building configuration...
Current configuration: 151 bytes

interface GigabitEthernet 0/3
 reverse-path
 ip address 10.1.0.38 255.255.255.252
 ip ospf network point-to-point
 ip ospf cost 5
 ip nat inside
```

（5）使用"show flow-control"命令，查看流控信息，如下所示。

```
EG1#show flow-control
flow-control Gi0/1
 channel-tree inbound
  no auto-pir enable
  !
  channel-group root parent null cir 1000000 pir 1000000 pri 4 fifo
  channel-group ssh parent root pir 100000 pri 4 per-net per-pir 1000 limit 2000
  channel-default root
```

```
!
 channel-tree outbound
  no auto-pir enable
  !
  channel-group root parent null cir 1000000 pir 1000000 pri 4 fifo
  channel-group ssh parent root pir 100000 pri 4 per-net per-pir 1000 limit 2000
  channel-default root
 !
 flow-rule 1 app-group SSH time-range any
 flow-rule 1 action pass in-channel ssh out-channel ssh comment ssh
!
flow-control Gi0/2
 channel-tree inbound
  no auto-pir enable
  !
  channel-group root parent null cir 1000000 pir 1000000 pri 4 fifo
  channel-group ssh parent root pir 100000 pri 4 per-net per-pir 1000 limit 2000
  channel-default root
 !
 channel-tree outbound
  no auto-pir enable
  !
  channel-group root parent null cir 1000000 pir 1000000 pri 4 fifo
  channel-group ssh parent root pir 100000 pri 4 per-net per-pir 1000 limit 2000
  channel-default root
 !
 flow-rule 1 app-group SSH time-range any
 flow-rule 1 action pass in-channel ssh out-channel ssh comment ssh
```

（6）使用 "show run | inc route-auto-choose" 命令，查看配置信息，如下所示。

```
EG1#show run | inc route-auto-choose
route-auto-choose cnc GigabitEthernet 0/1 196.1.0.2
route-auto-choose cernet GigabitEthernet 0/2 197.1.0.2
```

（7）使用 "show content-policy" 命令，查看策略信息，如下所示。

```
EG1#show content-policy
content-policy _TOP_PRIORITY
 (active)app-rule 200 time-range any app-group Block_Group action deny audit
 (active)app-rule 197 time-range any app-group Block_Group action deny audit vpn
 (active)url-rule 1000 url-object illegal time-range any action deny audit comment

content-policy p2p
 (active)app-rule 1 time-range p2p app-group P2P0ction deny audit
```

（8）使用 "show time-range" 命令，查看时间段信息，如下所示。

```
EG1#show time-range

time-range entry: any (active)
  periodic Daily 0:00 to 23:59

time-range entry: luyou (inactive)
  periodic Daily 16:00 to 22:00

time-range entry: p2p (active)
  periodic Weekdays 9:00 to 17:00
```

14. 上海本部校区出口设备 EG2 配置参考

（1）使用 "show ip int brief" 命令，查看接口摘要信息，如下所示。

```
EG2#show ip int brief
Interface              IP-Address(Pri)    IP-Address(Sec)   Status
GigabitEthernet 0/0    10.1.0.10/30       no address        up
GigabitEthernet 0/1    196.1.0.2/24       no address        up
GigabitEthernet 0/2    197.1.0.2/24       no address        up
GigabitEthernet 0/3    10.1.0.42/30       no address        up
GigabitEthernet 0/4    172.16.0.202/24    no address        up
Loopback 0             11.1.0.12/32       no address        up
SSLVPN 0               no address         no address        down
SSLVPN 1               no address         no address        down
```

（2）使用 "show url-class user-cfg" 命令，查看 URL 信息，如下所示。

```
EG2#show url-class user-cfg
url-class:forbidClass
       url: 196.1.0.2
```

（3）使用 "show app route" 命令，查看 App 路由信息，如下所示。

```
EG1#show app route
CLASS              SRC-GRP    DST-GRP                          INTERFACE(GROUP)
-----              -------    -------                          ----------------
P2P0               any        any                              GigabitEthernet 0/2
```

任务 ③⑦ 综合实训 2

【测试验证】

说明：因为使用一台设备进行测试，所以测试中的物理地址、本地连接 IPv6 地址等信息会出现一样的情况。

（1）测试计算机 PC1 连接上海本部无线获取 IP 地址，使用 "ipconfig/all" 命令查看设备信息，查看结果如下所示。

```
C:\>ipconfig/all
    连接特定的 DNS 后缀 . . . . . . . :
    描述 . . . . . . . . . . . . . . : Intel(R) Centrino(R) Advanced-N 6205
    物理地址 . . . . . . . . . . . . : 08-11-96-6E-1A-F8
    DHCP 已启用 . . . . . . . . . . : 是
    自动配置已启用 . . . . . . . . . : 是
    本地连接 IPv6 地址 . . . . . . . : fe80::d400:af4f:f101:7e15%8(首选)
    IPv4 地址 . . . . . . . . . . . : 192.1.60.1(首选)
    子网掩码 . . . . . . . . . . . . : 255.255.255.0
    获得租约的时间   . . . . . . . . : 2018 年 1 月 12 日 19:59:48
    租约过期的时间   . . . . . . . . : 2018 年 1 月 13 日 19:59:48
    默认网关 . . . . . . . . . . . . : 192.1.60.254
    DHCP 服务器 . . . . . . . . . . : 192.1.60.254
    DHCPv6 IAID . . . . . . . . . . : 117969302
    DHCPv6 客户端 DUID . . . . . . . : 00-01-00-01-21-96-BB-B2-10-1F-74-F3-C7-EE
    DNS 服务器 . . . . . . . . . . . : fec0:0:0:ffff::1%1
                                        fec0:0:0:ffff::2%1
                                        fec0:0:0:ffff::3%1
    TCPIP 上的 NetBIOS   . . . . . . : 已启用
```

（2）在测试计算机 PC1 上使用 "tracert 197.1.0.2" 命令测试路由跳数。

```
C:\Users\asin>tracert 197.1.0.2
```

可以发现，通过最多 30 个跃点跟踪到 197.1.0.2 的路由。

```
  1    16 ms     1 ms     1 ms   192.1.60.252
  2     4 ms     2 ms     3 ms   192.1.100.253
  3     3 ms     2 ms     2 ms   197.1.0.2
```

（3）在测试计算机 PC2 连接北京分校无线获取 IP 地址，使用 "ipconfig/all" 命令查看无线局域网适配器 WLAN 信息。

```
C:\>ipconfig/all

    连接特定的 DNS 后缀 . . . . . . . :
    描述 . . . . . . . . . . . . . . : Intel(R) Centrino(R) Advanced-N 6205
    物理地址 . . . . . . . . . . . . : 08-11-96-6E-1A-F8
    DHCP 已启用 . . . . . . . . . . : 是
    自动配置已启用 . . . . . . . . . : 是
    本地连接 IPv6 地址 . . . . . . . : fe80::d400:af4f:f101:7e15%8(首选)
    IPv4 地址 . . . . . . . . . . . : 194.1.20.4(首选)
```

多层交换技术（实践篇）

```
子网掩码  . . . . . . . . . . . . : 255.255.255.0
获得租约的时间  . . . . . . . . . : 2018 年 1 月 12 日 20:00:37
租约过期的时间  . . . . . . . . . : 2018 年 1 月 13 日 20:00:37
默认网关. . . . . . . . . . . . . : 194.1.20.254
DHCP 服务器  . . . . . . . . . . : 194.1.20.254
DHCPv6 IAID . . . . . . . . . . . : 117969302
DHCPv6 客户端 DUID  . . . . . . . : 00-01-00-01-21-96-BB-B2-10-1F-74-F3-C7-EE
DNS 服务器  . . . . . . . . . . . : fec0:0:0:ffff::1%1
                                    fec0:0:0:ffff::2%1
                                    fec0:0:0:ffff::3%1
TCPIP 上的 NetBIOS  . . . . . . . : 已启用
```

（4）在测试计算机 PC3 连接广州分校无线获取 IP 地址，使用"ipconfig/all"命令查看无线局域网适配器 WLAN 信息。

```
C:\>ipconfig/all

连接特定的 DNS 后缀 . . . . . . . :
描述. . . . . . . . . . . . . . . : Intel(R) Centrino(R) Advanced-N 6205
物理地址. . . . . . . . . . . . . : 08-11-96-6E-1A-F8
DHCP 已启用 . . . . . . . . . . . : 是
自动配置已启用. . . . . . . . . . : 是
本地连接 IPv6 地址. . . . . . . . : fe80::d400:af4f:f101:7e15%8(首选)
IPv4 地址 . . . . . . . . . . . . : 195.1.20.3(首选)
子网掩码  . . . . . . . . . . . . : 255.255.255.0
获得租约的时间  . . . . . . . . . : 2018 年 1 月 12 日 20:01:12
租约过期的时间  . . . . . . . . . : 2018 年 1 月 13 日 20:01:12
默认网关. . . . . . . . . . . . . : 195.1.20.254
DHCP 服务器  . . . . . . . . . . : 195.1.20.254
DHCPv6 IAID . . . . . . . . . . . : 117969302
DHCPv6 客户端 DUID  . . . . . . . : 00-01-00-01-21-96-BB-B2-10-1F-74-F3-C7-EE
DNS 服务器  . . . . . . . . . . . : fec0:0:0:ffff::1%1
                                    fec0:0:0:ffff::2%1
                                    fec0:0:0:ffff::3%1
TCPIP 上的 NetBIOS  . . . . . . . : 已启用
```

（5）在测试计算机 PC3 上使用"tracert 197.1.0.1"命令测试路由跳数。

```
C:\Users\asin>tracert 197.1.0.1
```

可以发现，通过最多 30 个跃点跟踪到 197.1.0.1 的路由。

```
  1     4 ms     3 ms     4 ms   195.1.20.254
  2     1 ms     1 ms     1 ms   10.1.0.29
  3    87 ms    45 ms    36 ms   10.1.0.22
  4    43 ms    42 ms    44 ms   10.1.0.13
  5    49 ms    48 ms    49 ms   10.1.0.10
  6    45 ms    45 ms    45 ms   197.1.0.1
```